Housing in Turkey

This book presents the major features of the path that the Turkish housing system has followed since 2000. Its primary focus is to build an understanding of housing in Turkey from the policy, planning, and implementation perspectives in the 21st century, interwoven with the effects of neoliberalism.

It investigates the social, spatial, and economic outcomes of the shift in philosophy and behaviour by the government regarding housing. The book discusses failures in housing outcomes as government failures, incorrect or inefficient regulations, lack of regulations, and lack of monitoring of the policy outcomes. Chapters on the housing-economy relationship, financialization and indebtedness, housing market experiences based on case studies, and the housing policy provide the reader with an opportunity to observe different outcomes in a world where housing challenges and issues are similar.

This book will be of interest to urban planners, political scientists, and sociologists, as well as undergraduate/graduate students and housing sector experts all over the world who are interested in the various dimensions of the housing problem.

Ö. Burcu Özdemir Sarı is an Associate Professor at the Department of City and Regional Planning, Middle East Technical University (METU), Turkey. She earned her PhD in Spatial Planning at TU Dortmund, Germany. She conducted research projects about housing affordability, residential vacancy, and property rights. She is currently teaching urban economics and housing. Her research interests are urban studies, housing economics, housing policy, and residential reinvestment. She is the editor of the Housing Book Series launched by the Institute of Urban Studies (Turkey).

Esma Aksoy Khurami is currently an Assistant Professor at the Faculty of Architecture, the Department of City and Regional Planning at Muğla Sıtkı Koçman University. She earned her bachelor's (2014), MSc (2017), and PhD (2021) degrees from Middle East Technical University. She received several awards with her MSc thesis, including the Nathaniel Lichfield Award by Regional Studies Association in 2017. Her research interests focus on low-income households, housing affordability, homeownership, quantitative methods, and techniques in housing and planning studies.

Nil Uzun is a Professor at the Department of City and Regional Planning, Middle East Technical University (METU), Turkey. She earned her PhD degree in Urban Geography at Utrecht University in 2001 with her dissertation on gentrification in İstanbul. She participated in and conducted research projects about urban transformation. Along with teaching urbanization and urban sociology, she is doing research mainly on gentrification and residential transformation.

Housing in Turkey
Policy, Planning, Practice

Edited by
Ö. Burcu Özdemir Sarı, Esma Aksoy
Khurami and Nil Uzun

LONDON AND NEW YORK

First published 2022
by Routledge
4 Park Square, Milton Park, Abingdon, Oxon OX14 4RN

and by Routledge
605 Third Avenue, New York, NY 10158

Routledge is an imprint of the Taylor & Francis Group, an informa
business

British Library Cataloguing-in-Publication Data
A catalogue record for this book is available from the British Library

Library of Congress Cataloging-in-Publication Data
Names: Khurami, Esma Aksoy, editor. | Sari, Ö. Burcu Özdemir,
editor. | Uzun, Nil, editor.
Title: Housing in Turkey : policy, planning and practice / edited by
Ö. Burcu Özdemir Sari, Esma Aksoy Khurami and Nil Uzun.
Description: Abingdon, Oxon ; New York, NY : Routledge, 2022. |
Includes bibliographical references and index.
Identifiers: LCCN 2021051429 (print) | LCCN 2021051430 (ebook) |
ISBN 9781032003269 (hardback) | ISBN 9781032003276 (paperback) |
ISBN 9781003173670 (ebook)
Subjects: LCSH: Housing--Turkey. | Housing policy--Turkey.
Classification: LCC HD7358.25.A3 H68 2022 (print) |
LCC HD7358.25.A3 (ebook) | DDC 363.509561--dc23/eng/20211020
LC record available at https://lccn.loc.gov/2021051429
LC ebook record available at https://lccn.loc.gov/2021051430

ISBN: 978-1-032-00326-9 (hbk)
ISBN: 978-1-032-00327-6 (pbk)
ISBN: 978-1-003-17367-0 (ebk)

DOI: 10.4324/9781003173670

Typeset in Times New Roman
by KnowledgeWorks Global Ltd.

Contents

Figures

Tables

Contributors

Esma Aksoy Khurami is currently an Assistant Professor at the Faculty of Architecture, the Department of City and Regional Planning at Muğla Sıtkı Koçman University. She earned her bachelor's (2014), MSc (2017), and PhD (2021) degrees from Middle East Technical University. She received several awards with her MSc thesis, including the Nathaniel Lichfield Award by Regional Studies Association in 2017. Her research interests focus on low-income households, housing affordability, homeownership, quantitative methods, and techniques in housing and planning studies.

Leyla Alkan Gökler received a bachelor's degree in City and Regional Planning and a master's degree in Economics from Middle East Technical University. She completed her PhD on Tenure Choice and Demand for Homeownership at the same university in 2011. Currently, she is an Associate Professor in the Department of City and Regional Planning at Gazi University. Her major research interests include housing, urban economics, real estate economics, and urban property development.

Zeynep Arslan Taç is a PhD candidate in Social and Political Sciences at Marmara University. Her research interests are the meaning of home, housing policy, and financialization of housing. She received her BA in Political Science at İ.D. Bilkent University (2010) and her MSc in Urban Policy Planning and Local Governments at Middle East Technical University (2014). She has been working as a research assistant at the Department of Political Science and Public Administration at Marmara University since 2012.

Ebru Kamacı Karahan is an Assistant Professor at Bursa Technical University, Turkey. She completed her PhD dissertation at Middle East Technical University entitled "Re-reading Urbanization Experience of Istanbul through Changing Residential Mobility Behaviour of Households". She researches housing, with a particular focus on intra-urban mobility, residential satisfaction, and segregation.

Esra Alkim Karaagac is a researcher and a Geography PhD candidate at the University of Waterloo. Her research interests include urban economic restructuring, financialization of housing, and feminist methodologies in economic geography. For her dissertation, she examines the role of indebtedness on the conduct of citizens by investigating the state-led housing provision in Turkey. Alkim studied for a double major in City and Regional Planning and Sociology at the Middle East Technical University. She holds a master's degree in City Design and Social Science from the London School of Economics and Political Science as a Jean Monnet scholar.

H. Kübra Kıvrak works as a city planner at a private company. She earned her bachelor's degree in City and Regional Planning and a minor degree in Geographic Information Systems and Remote Sensing at Middle East Technical University (METU) in 2015. She earned her master's degree in City Planning at METU in 2019 with her evaluation on residential vacancy patterns in Gemlik/Bursa. Her research interests are high housing production and its outcomes in Turkey, housing vacancy, and different residential vacancy patterns.

Yelda Kızıldağ Özdemirli received her BA, MSC, and PhD degrees from Middle East Technical University. She has held research, visiting research, and teaching positions in Middle East Technical University, UC Berkeley, Selçuk University, and Konya Technical University. She is passionate about conducting research on housing policy implications of housing needs of households. Her research spans through a variety of aspects of housing; namely quality of housing, urban redevelopment, community development, management of housing, informal housing, housing for the elderly, and housing policy.

Tuğba Kütük received her bachelor's degree (2018) from the City and Regional Planning Department, Faculty of Architecture, Middle East Technical University. She has been conducting her master's research in the Department of City and Regional Planning, Faculty of Architecture, Gazi University. She has been working in the Department of City and Regional Planning at Gazi University as a research assistant since 2019. Her research interests include urban studies, urban economics, and housing studies.

Ö. Burcu Özdemir Sarı is an Associate Professor at the Department of City and Regional Planning, Middle East Technical University (METU), Turkey. She earned her PhD in Spatial Planning at TU Dortmund, Germany. She conducted research projects about housing affordability, residential vacancy, and property rights. She is currently teaching urban economics and housing. Her research interests are urban studies, housing economics, housing policy, and residential reinvestment. She is the editor of the Housing Book Series launched by the Institute of Urban Studies (Turkey).

Tanyel Özelçi Eceral received her B.CP. (1990), M.RP (1994) and her PhD (2004) from City and Regional Planning Department, Faculty of Architecture, Middle East Technical University. She had worked for the Ministry of Culture and for the South-eastern Anatolia Project Regional Development Administration. She had been working in the Department of City and Regional Planning at Gazi University as a Faculty Member between the years 1995 and 2001, and has been working as an academic since 2003. She had been the Vice-Chair of the Department and Member of the Faculty Board. Her research interests are regional studies, city and regional planning, urban economics.

G. Pelin Sarıoğlu Erdoğdu graduated from the Department of City and Regional Planning at Middle East Technical University. Received her Master's degree in the same university in 2003. For her PhD, she studied at the University of Groningen in The Netherlands as an Ubbo Emmius PhD scholarship holder. Her PhD dissertation was published as a book in 2014. Holds associate professorship title since 2017. Currently works in the Department of Architecture at Doğuş University in Istanbul. Her research interests are housing research, housing tenure studies, basic design, and design education.

Aynur Uluç received her bachelor's degree in City and Regional Planning (2011) and master's degree in Restoration and Conservation of Historic Sites and Monuments (2014) from Middle East Technical University (METU). She continues her PhD in City and Regional Planning at METU and currently works as a Research Assistant at the same university. Major research interests include urban conservation planning, risk management, and mitigation planning for the conservation of cultural heritage.

Nil Uzun is a Professor at the Department of City and Regional Planning, Middle East Technical University (METU), Turkey. She earned her PhD degree in Urban Geography at Utrecht University in 2001 with her dissertation on gentrification in İstanbul. She participated in and conducted research projects about urban transformation. Along with teaching urbanization and urban sociology, she is doing research mainly on gentrification and residential transformation.

Preface

This book project was launched just a few months before the Covid-19 outbreak. During the initial stage of the project, we were hoping to establish a network among housing researchers in the country. The book would be the outcome of a workshop where scholars in the field were invited to present their recent research. However, Covid-19 has changed everything very fast; meetings have been cancelled, universities have shifted to online education, lockdowns have become a regular part of our lives, and most people in the world have started to work where they live. As a result, not only have the planned processes for this book project had to change, but also the focus of our research, which is housing issues, has been affected. The meaning of home in our minds and housing preferences in society have changed dramatically during this time. New policy measures have emerged to tackle the pandemic. It has become clear that housing is an indispensable pillar for the well-being of society, and housing problems require special attention in terms of policy and planning. The pandemic has shown us once again, as in the periods of severe economic crisis, the role of housing in the economy, the significance of having a decent and affordable home, and the vital role of access to open and green spaces in housing areas.

Housing in Turkey: Policy, Planning, Practice has evolved as a result of the above-mentioned chaotic conditions unleashed by the pandemic. The book is as comprehensive as it can be to discuss different dimensions of housing in Turkey as it is necessary for educational purposes as well as for international readers interested in the Turkish housing market. It is a fact that any study in the field of housing would remain incomplete since this field is associated with various disciplines that are not possible to cover in a single book. Nevertheless, we have attempted to explore several dimensions of the field as extensively as possible. Although there are several comprehensive international studies in the housing field, books discussing various dimensions of housing in the Turkish context are absent both nationally and internationally. We hope that this book will be useful for undergraduate/graduate students, housing sector experts, researchers, and academics all over the world.

Although we experienced some difficulties during the preparation of the book due to the conditions imposed by the pandemic, several valuable people assisted us to ensure the process ran as smoothly as possible. The editors would like to thank the authors for their committed contributions, particularly for their patience in responding to the reviewers' and editors' demanding comments. It should be noted that all authors (including the editors) of the book are female. Of course, this is not a deliberate choice but a coincidence. It is a fact that the domestic burden borne by women has increased during the pandemic. Considering this fact, we believe that under highly difficult conditions, the authors and editors have done a great job. We would also like to thank the following experts in the housing field for reviewing initial drafts of the chapters and providing invaluable guidance and suggestions: Prof. Dr. Nihan Özdemir Sönmez, Assoc. Prof. Dr. Zeynep Günay, Asst. Prof. Dr. Melih Yeşilbağ, and Asst. Prof. Dr. Ebru Kamacı Karahan. Last but not least, we would like to thank our families for their considerable patience and indulgence throughout the preparation of the book. Special thanks go to our children, Ece Sarı, Mete Sarı, Selin Uzun, and Orhun Uzun.

To our mothers, Gönül Özdemir, Sunay Aksoy, and Işın Duruöz. They have played the biggest part in making us who we are today.

Ankara, Turkey
Ö. Burcu Özdemir Sarı
Esma Aksoy Khurami
Nil Uzun

1 Introduction

*Ö. Burcu Özdemir Sarı, Esma Aksoy
Khurami, and Nil Uzun*

The Changing Context of Housing Since 2000

The first two decades of the twenty-first century witnessed many events such as economic crises, disasters, international migration, and a pandemic that required rethinking the role and meaning of housing in terms of society and economy. The deepening of the existing social and spatial inequalities resulting from these developments has turned the attention of governments in many countries to housing research. Housing policies were renewed and updated in line with the findings of these studies. Not only did housing policies undergo a significant transformation, but also the functions and meaning of housing changed in this period. In addition to its most basic function as a shelter, housing has many other functions in the economic, social, cultural, and political fields in today's world. Housing contributes to the economy as a sector, affects economic activity due to its strong relations with other sectors, and determines the mobility dynamics in the labour market. In addition, it affects the formation of human capital and the life chances of new generations, shapes the consumption and investment behaviour of the society through the housing wealth of the households (Regeneris Consulting and Oxford Economics, 2010), and becomes a cultural asset that transfers the architectural and aesthetic features of the past to the present. Furthermore, with the spread of Covid-19, it has become an essential health determinant (Housing Research Collaborative, 2020), and its function as a workplace has become more important than ever with remote working requirements during the lockdown.

The events that have had worldwide effects so far this century, such as the global financial crisis of 2008, the Covid-19 pandemic, and the 'global housing affordability crisis' (Wetzstein, 2017; Kallergis et al., 2018; Coupe, 2020), show that countries have a lot to learn from each other. Of course, this does not mean that countries' housing policies will display convergence in response to global challenges or that solutions that work in one context can be applied in others. It should be acknowledged that path dependence is significant in housing policy (Gurran and Bramley, 2017). However, it is still possible to benefit from each other's housing experiences, and sometimes even negative

DOI: 10.4324/9781003173670-1

experiences could also be useful. In that sense, this book provides an opportunity for the reader to observe different outcomes in a world where housing challenges and issues are similar. One can argue that housing problems are different in developing and developed parts of the world. Nevertheless, global challenges affect each and every country. For instance, during the Covid-19 pandemic, loss of livelihood has resulted in the increased risk of evictions not only in lower- and middle-income countries but also in high-income ones (Housing Research Collaborative, 2020). Moreover, as Gurran and Bramley (2017: 3) note, "Whilst problems of slum housing, overcrowding and inadequate sanitation remain concentrated in the developing world, severe affordability problems plague the so-called richest nations as well".

Today, an increasing proportion of the world's population is unable to access decent, adequate, affordable housing and sustainable living environments. Many European countries have experienced residualization of the social housing sector recently with a continuing shift to owner-occupation and gradually less direct intervention on the part of the state. The financialization of housing emerged as a result, and investment and wealth functions of housing have become prominent. Turkey, however, has followed a different path this century with social housing in the form of subsidized owner-occupation having been initiated, and the state has become a direct actor in housing provision. Housing in Turkey was already commodified due to the lack of welfare state measures and comprehensive housing policies and through the historical promotion of (single/multiple) homeownership. However, in the first two decades of this century, the Turkish experience has transformed into a "state-orchestrated financialization of housing" (Yeşilbağ, 2020). In this context, this book presents an interesting example with increasingly high housing output but still increasing house prices and the deepening affordability gap among the different segments of society. High rates of homeownership, the significant market share of the private rented sector, the total absence of a social rented sector, a recently developed mortgage finance system, and gradually increasing state involvement as a direct actor in housing production are the other prominent features of the Turkish housing system.

Housing in Turkey in the Twenty-First Century

The main theme of this book is housing in Turkey in the twenty-first century, in view of the fact that since the early 2000s, housing has become one of the central political, economic, and social issues in the country. In 2002, the Justice and Development Party (AKP) came to power and continues to rule almost 20 years later. Therefore, this book mainly reflects policy, planning, and practice under the implementations of AKP governments. During these 20 years, the role of the housing and construction industry in economic development and housing as a financial investment have become the major reasons behind the supremacy of housing in Turkey. The result of the emphasis that has been placed on the housing and construction sector is an extraordinary surge in housing output across the country. The annual housing starts reached a record high level in 2014 with one million new dwelling units. A new record was broken in 2017 with 1.4 million new housing starts. Of course, these figures far exceeded the housing requirement of the society.

The housing boom in Turkey is observed in urban areas, in the lands allocated by the planning authorities for new development, and in existing squatter housing areas and historical parts of cities. Furthermore, coasts, flood plains, plateaus, pastures, forests, and even conservation areas have seen their share of this construction boom. With the neoliberal policies gaining momentum in the 2000s, attempts to increase construction activities in the country emerged as large-scale infrastructure and housing projects, a countrywide housing programme, and associated social housing programme (subsidized owner-occupation) by the public sector, and a move towards urban transformation. A broad range of incentives was also adopted to support the continuity of the construction sector's activities, such as interest rate cuts on mortgages and housing sales campaigns led by the public sector. These attempts turned the country into a construction site and marked the beginning of the 'bulldozer era' for the areas with locational advantages, such as the squatter housing areas and the deteriorated historical parts of the cities.

It is hardly surprising to see housing-related news in newspapers and on television almost every day given the circumstances mentioned above. The demolition of squatter housing areas, problems or opportunities in urban transformation projects, severe damages experienced in hazard-prone settlements, rising housing prices, fluctuations in housing sales, state-led housing campaigns, changes in mortgage interest rates, and many more are now very familiar issues for ordinary citizens. Interestingly, very little research has been done to investigate the social, spatial, and economic outcomes of this shift in philosophy and behaviour by the government regarding housing. Just like the government and public administrations that are not interested in the outcomes of the adopted policies, researchers' interest in this area is quite limited. The lack of comprehensive and sufficient data sources for housing studies can be considered an acceptable reason for this situation. Nevertheless, this book is proof that it is possible to study the country's housing problems, albeit with limited data. In that context, the purpose of this book is to capture several dimensions of housing as well as possible, such as housing as a subject of research, policy, and planning as well as housing as a commodity, a right for all segments of the society, and an area of government intervention.

The Perspective of the Book

It is evident that any study in the field of housing would remain incomplete since this field is at the intersection of various disciplines that are not possible to cover in a single book. The book's editors and contributors have a background in urban and regional planning and public policy. However, urban and regional planning is a broad discipline that allows for specializations like urban sociology, urban economics, and urban policy. Therefore, this book fundamentally has a public policy and planning perspective touching on sociology and economics from time to time. The spatial dimension is emphasized in most of the chapters since the book includes several case studies. Knowing that planning and policy are one side of the coin, while the market is the other in market-based economies, the book also considers the role of the market in the production of housing and the built environment.

However, failures of housing markets in terms of quality, quantity, and affordability are mostly considered failures of urban planning and public policy in the context of the book. This is because the decision to allocate land for new development and thereby the amount of new housebuilding and location choice; the size of urban plots and development density, so size and type of housing; and the urban transformation decision are determined by the planning authority. These decisions also have effects on the price of housing. In addition, public policy is responsible for countrywide housing strategies and measures. This means, though limited in some aspects, planning and policy exert significant control over the production of housing and living environments. Without the sole reliance of housing policy on homeownership and the lack of alternative investment tools for households in volatile economic conditions, homeownership would not be a safe haven for households to accumulate, secure, and increase household wealth. In that case, the rental stock would be one that is developed purposefully rather than being a by-product of homeownership in the free market environment. Alternately, if the state had not fuelled the construction sector in order to stimulate the economy, excess production, high residential vacancy rates, and uneven distribution of housing output would not have been the results of the free market processes. In that sense, most of the chapters in the book discuss failures in housing outcomes not only as market inefficiencies but most of the time as government failures, incorrect and/or inefficient regulations, lack of regulations, and lack of monitoring of the policy outcomes.

Structure of the Book

As Clark (2021: xvi) mentions in his recent book, there are different ways to study housing, and the field is evolving via different approaches emphasizing different components of the "puzzle of housing". This is also true for this book in the sense that an array of views is reflected in different parts of the study. That is the reason for providing a general framework for the study in the first part of the book and leaving the conceptual/theoretical framework and relevant literature to be discussed in each of the chapters. This also helps to keep each chapter as a standalone section discussing a theme coherently and also contributing to the book's general discussion. In that sense, Part I presents the framework of the study; Part II evaluates the role of housing for the economy, the housing–economy relationship, and the consequences of this relationship in terms of financialization and indebtedness. Part III comprises chapters discussing housing market experiences based on case studies, with an empirical focus specifically. The final part, Part IV, brings the policy-related chapters together and presents discussions on the implications of housing policy or lack of it.

In Part I, Chapters 2 to 5 cover the fundamentals of housing in Turkey. In this context, Chapter 2 presents an overview of the Turkish housing system, including the tenure structure and related tendencies, physical features of the existing stock, the state's changing role in housing provision and housing production, and the role of the planning system in the housing market. Chapter 3 evaluates the policy dimension and reviews the political and economic background of housing policies before and after 2000. The philosophical change in government actions in the post-2000 period is emphasized in the chapter, where housing comes to the forefront as a major social and economic policy area with 'countrywide housing production' and 'urban transformation' programmes. In Chapter 4, the housing and living conditions of households are compared for 2006 and 2018 in terms of poverty status, material deprivation, housing-related living conditions, adequacy of current income, and the burden of housing and non-housing expenditures. Chapter 5 highlights the limitations that housing researchers face. It also discusses the availability of data for housing research in Turkey and emphasizes the need for robust and continuous data for better research. These four chapters provide a framework for the discussions found in the other chapters of the book.

In Part II, Chapters 6 to 8 discuss housing, the economy, and financialization. There is an emerging body of literature about the financialization of housing in Turkey. However, there is a need to elaborate on the link between housing and the economy, public policy, and the socio-spatial dimension. Chapter 6 shows that the country's economic environment directly affects housing policies, and the housing policies applied to date have been unable to solve the existing problems and have actually contributed to the creation of further problems. Chapter 7 mainly focusses on the changing role of the state from a regulator to an active financial market actor and highlights that Turkey has been experiencing state-led financialization in the housing market. Moreover, Chapter 7 identifies Turkey as an example of the variegated structure of financialization of housing in the Global South. In Chapter 8, the Housing Development Administration's (TOKİ) mass housing programmes for low-income groups are examined, focussing on everyday negotiations of finance and debts in housing estates. The chapter argues that households, being indebted to the state for their homes, become financial subjects through the credit system of state-owned banks. In other words, Chapter 6 examines housing from an economic perspective, whereas

Chapter 7 has a more political focus. Chapter 8 deals more with everyday life and indebtedness in relation to financialization.

In Part III, Chapters 9 to 12 discuss housing market experiences based on case studies. Due to the lack of countrywide comprehensive and robust data on housing, the case study approach is inevitable for the study of topics such as residential vacancies, housing submarkets, gated communities, and traditional housing stock. In that context, Chapter 9 displays the extent of the surplus housing stock in the country, employing official statistical data, and elaborates its findings through a case study in the Gemlik District of Bursa Province. Based on the case study, the chapter shows that vacancy follows different patterns in different housing submarkets. Moreover, vacant stock in the country is not only owned by the construction firms but also by individuals, an issue that current measures and strategies neglect. Chapter 10, utilizing data from advertisements for residential real estate in Ankara, empirically investigates the housing submarkets. The chapter reveals that spatially defined submarkets are more relevant than socio-economically defined submarkets for explaining the house price structure. Furthermore, the study highlights that the housing price structure in each submarket is dependent on different variables. Considering the housing policies in Turkey, which are enacted at the country level, the findings of Chapter 10 emphasize that 'one size does not fit all'. Chapter 11 conducts an exploratory qualitative field study to examine a widespread phenomenon, gated communities. The city of Bursa serves as a case study area. The chapter focusses on the internal dynamics of living in gated communities in different locations and with different development histories. The study reports that the gated communities of the 1990s are different from those of the 2000s in terms of residential types, development history, and location preferences. The chapter concludes that with the urban transformation movement in the country living in such a community is no longer unique to upper-status groups. Finally, Chapter 12 focusses on traditional housing stock, employing a case study in the Kaleiçi area of Antalya, and analyses the challenges faced by the residents of housing units that are listed as cultural property during their repair, maintenance, and restoration. The findings of the chapter show that legal limitations and bureaucratic hurdles affect the use and condition of housing units in traditional areas, and due to the lack of financial support, the high costs of repair, maintenance, and restoration prevent the sustainability of these areas.

In Part IV, Chapters 13 to 15 evaluate the implications of housing policy (or lack thereof). Chapter 13 focusses on housing for ageing populations and investigates the potentials and limitations of the housing stock and housing policy in the country to cope with the needs of the ageing population. The study results reveal that an ageing population might result in a housing provision crisis for the elderly in Turkey in terms of the number, affordability, and quality. Furthermore, the results of the chapter indicate that Turkey needs to initiate policy and strategies for developing housing designated primarily for the elderly but also to re-evaluate the conditions of the existing stock and neighbourhoods in terms of 'ageing in place'. Chapter 14 empirically investigates the housing and transport expenditure burden of tenant households in the country. At the same time, the chapter also reviews the fundamental issues related to the private rented sector in Turkey. As the chapter shows, the provision of rental units has never been on the Turkish governments' agenda, and the private rented sector emerged as a by-product of homeownership. In the chapter, it is argued that both the sector and the tenant households have been left to their fate in the free market environment. In this context, the chapter investigates the cost items affecting tenant households' housing and transport burden. The findings of the study reveal that the ratio of cost-burdened households due to housing and transport expenditures increases with movement on the income scale from higher to lower incomes. Moreover, it is the share of housing rent and private transport expenditures in households' budgets that create the main differences between tenants having a cost burden and those with no burden. Finally, Chapter 15 discusses urban transformation policies in twenty-first-century Turkey, focussing on state-led gentrification. As mentioned in the chapter, neoliberal economic policies generated new political and economic actors, leading to new forms of socio-spatial segregation. The focus of this chapter is on urban transformation policies, which have been dominant in the last two decades. The general discussion about the policies and the urban transformation project presented in this chapter reveals that state-led gentrification has been the prevailing urban transformation policy.

References

Clark WAV (2021) *Advanced introduction to housing studies.* Cheltenham: Edward Elgar Publishing.

Coupe T (2020) How global is the affordable housing crisis? *International Journal of Housing Markets and Analysis*, 14, 429–445.

Gurran N, Bramley G (2017) *Urban planning and the housing market: International perspectives for policy and practice.* London: Palgrave Macmillan.

Housing Research Collaborative (2020) COVID Housing Policy Roundtable Report. Available at: https://housingresearchcollaborative.scarp.ubc.ca/files/2020/11/FinalReport_COVID-19-Global-Housing-Policies.pdf. Accessed 16 June 2021.

Kallergis A, Angel S and Liu Y et al. (2018) *Housing affordability in a global perspective.* Lincoln Institute of Land Policy. Available at: https://www.lincolninst.edu/pt-br/publications/working-papers/housing-affordability-global-perspective. Accessed 20 June 2021.

Regeneris Consulting and Oxford Economics (2010) *The role of housing in the economy: A final report.* London: HCA. Available at: https://www.oxfordeconomics.com/publication/open/224366. Accessed 16 June 2021.

Wetzstein S (2017) The global urban housing affordability crisis. *Urban Studies*, 54, 3159–3177.

Yeşilbağ M (2020) The state-orchestrated financialization of housing in Turkey. *Housing Policy Debate*, 30(4), 533–558.

Part I

Housing in Turkey

Framework of the Study

2 The Turkish Housing System
An Overview

Ö. Burcu Özdemir Sarı

Introduction

It is not easy to review a country's housing system within the length of a book chapter since multiple variables such as institutions, policies, actors, and processes and their multifaceted relationships with each other make up the housing system. Furthermore, considering that the housing problem in any country is multidimensional, a reductionist approach is inevitable when determining the topics to be covered in such a chapter. One of the best methods, in this case, would be to provide the essential information necessary for those unfamiliar with the housing system of the country focussed on. Therefore, this chapter aims to present an overview of the key features of the Turkish housing system and the recently experienced transformations in it to set the scene for the discussions contained in the rest of the book.

As a starting point, it is significant to highlight a number of main arguments regarding the nature of the housing problem in Turkey. Turkey is a market-based economy, which means both the free market and government intervention have roles in allocating resources. In this dual system, central and local authorities and the free market mechanism have some control capacity over the property markets. As a result, although a significant amount of housing output has been created in the last 20 years, (i) geographically uneven distribution of this output across the country, (ii) affordability differences between regions and tenures, (iii) shrinking of the private rented sector, and (iv) safety and quality of life problems constitute the major housing challenges in Turkey (Özdemir Sarı, 2019:169). However, within the length of this chapter, it is not possible to cover all of these issues in detail. Only the salient features of the housing problems and policy are mentioned here to keep the text as coherent and consistent as possible. Housing policy, along with its historical development, on the other hand, is discussed comprehensively in Chapter 3 of this book.

It is also necessary to reflect on the definition of a housing system to determine the scope of the study. According to Oxley and Haffner (2012:199), a housing system refers to the "combination of the housing markets, institutions, and policies within a country that together deliver housing". In other

DOI: 10.4324/9781003173670-3

words, the housing system is a much broader term than the housing market, sector, or policy. Van der Heijden (2013) reviews several definitions of the term and identifies supply, demand, and institutions as the main components, also acknowledging the significance of the past and the context in the development of the housing system. The present study, considering the definitions mentioned above in the literature, in the first section, overviews the institutional context and regulations concerning housing development. The planning framework and the changing housing agenda in Turkey are presented in this part. Secondly, to understand the supply side issues, significant features of the housing production and existing housing and house prices are examined. The section thereinafter elaborates on the demand side, focussing on the demographic trends, changing tenure structure, housing wealth, and affordability. The final section is devoted to the conclusions. Throughout this review, the role of past policies and the context-specific issues are also acknowledged.

The Institutional Context and the Regulation of Housing Development

Compared to other commodities and commodity markets, housing is one of the most tightly regulated spheres (Harsman and Quigley, 1991). In many countries, housing development is subject to various regulations such as building codes, environmental regulations, and land-use regulations (Schill, 2005). These regulations influence the amount of housing production, location decisions, urban form, and house prices (Gyourko and Molloy, 2015). In the Turkish context, housing is regulated basically through the urban planning system and direct/indirect government interventions. Urban planning's control capacity over the property markets is weakened by government interventions external to the planning system since these two institutions are not coordinated with each other. In most cases, these interventions outside the planning system reflect political concerns and judgments and negatively impact the implementation of planning decisions.

The Planning Framework

Arrangements concerning spatial planning in Turkey date back to the nineteenth century (Özdemir Sönmez, 2017; Özdemir, 2019); however, the institutionalization of urban planning took place with the first Development Law passed in 1956 (Law no. 6785). The fundamental structure of the current spatial planning system, on the other hand, is set by the 1985 Development Law (Law no. 3194). Currently, based on demographic trends and population projections, spatial plans determine the physical aspects of housing development, such as the location, amount, and density of new development and the need for intervention in the existing housing areas. Furthermore, there is sectoral (economic) planning that started in the early 1960s with the establishment of the State Planning Organization (SPO) in 1960. Five-Year Development Plans, covering different sectors of the economy, have been prepared to support national and regional development. These development plans guide public policy and nationwide investments. In the field of housing, this means major topics of discussion about housing at the macro level and the major lines of the housing policy are defined in the development plans.

Before the enactment of the 1985 Development Law, there was already a spatial planning tradition with planning legislation to guide master plans and implementation plans. The major change brought about by this law is the transfer of the authority over planning from the central government to the local authorities (Ozkan and Turk, 2016). With the 1985 Development Law, the power to create and approve spatial plans was given to local authorities. This development led to increasing housing production in the country through increased planned areas and developed land (Türel and Koç, 2015). The vast majority of land in Turkey is privately owned. However, plans are legally binding, and the land-use decisions of urban planning impose limitations on the use of private property rights based on the public interest criterion. It is obligatory to prepare plans for settlements with a population exceeding 10,000 people. If the population is below 10,000, then the municipal council decides whether to prepare a plan or not.

In the current urban planning system, there are various kinds of plans. The major plan types are categorized as spatial strategic plans, environmental plans, master plans, implementation plans, and urban design projects (Özdemir, 2019). There is a hierarchical relationship between different plan types, and the implementation and development control is the consequence of the plan approval process. Master plans determine strategies and decisions for different land-use types regarding the plan area, the population, and building densities in line with the physical layout provided with upper-level plans (Ersoy, 2011). They are usually prepared at 1/5000 scale, but in metropolitan areas, the scale may be 1/25,000. Regarding the housing areas, in addition to the densities, master plans define, if any, the boundaries of the areas to be conserved, reorganized, and rehabilitated. Furthermore, the areas and boundaries determined by other laws (special-purpose laws) such as squatter housing prevention areas, mass housing areas, urban transformation, and development project areas, and renewal areas are displayed.

The implementation rules and construction guidelines are covered in 1/1000 scale implementation plans (Ersoy, 2011). In accordance with master plans, implementation plans define building layout (e.g., attached/detached), density (e.g., floor area ratio), height and setback distance, and other related information. In the planned areas, two types of permissions are granted by local authorities for new development: construction and occupancy permits. In order to start the construction activity, obtaining a construction permit is a legal necessity. When the construction has been completed in accordance with the project accepted at the construction permit stage, an occupancy permit is granted upon an additional application to the relevant authority. Obtaining an occupancy permit is considered costly and burdensome and thus is usually avoided even though it is a legal requirement. There were many dwelling units occupied without occupancy permits in the country. However, this trend is changing with the enactment of the new Turkish Criminal Law (Law no. 5237) in 2004 and the Planned Areas Development Regulation in 2017. Usually, housing starts (construction permits) are used rather than completions (occupancy permits) to represent housing production levels in the country (Türel and Koç, 2015).

According to the Development Law, unauthorized development incurs penalties. However, it was not possible to prevent unauthorized development in the country for several reasons. In the early decades of the Turkish Republic, the country's resources were limited, and there were several priorities other than housing requiring resource allocation. In addition, urban areas had experienced a massive migration from rural areas starting in the 1950s. Consequently, immediate housing demand in urban areas, which the public policies had not met, was met through the free market mechanism in the form of squatter housing construction (particularly for rural migrants) and multi-unit, multi-owned blocks of flats (Özdemir Sarı, 2019). As a result, a dual structure emerged in the housing market consisting of legal and illegal housing stock.

Squatter housing made up a significant share of illegal housing markets before 2000. Through several Amnesty Laws, squatter housing, an entirely illegal undertaking at the construction stage, became a way to achieve legitimate urban property ownership (Balamir, 1996). It was estimated that there were nearly 2 million squatter houses in 2002, accommodating 27% of the population in urban areas (Keleş, 2002). Since 2003, deliberate efforts by subsequent governments have resulted in massive clearance and redevelopment in squatter housing areas (Özdemir Sarı and Aksoy Khurami, 2018). Nevertheless, unauthorized housing construction is not limited to squatters. Buildings having construction permits but occupied without occupancy permits are also considered unauthorized. On the legal side of the market, the construction of blocks of flats became the predominant form of construction in the country with the enactment of the Flat Ownership Law in 1965 (Özdemir Sarı, 2019).

In the long run, it became clear that the 1985 Development Law had resulted in the fragmentation of planning authority (Duyguluer, 2006). The special-purpose laws, which appeared on the agenda as a result of the Development Law, authorized a number of central government units to prepare a plan in their fields of implementation (Ozkan and Turk, 2016). Thereby, these special-purpose laws increased the planning practices outside the planning system, and these uncoordinated interventions continue to weaken the effectiveness of urban planning. As of 2011, almost 30 public institutions are involved in plan preparation in their particular fields (Ersoy, 2011). Despite the attempts to empower local authorities in the 1980s, planning still has a highly centralized structure in Turkey (Özdemir Sarı et al., 2019).

The Changing Housing Agenda in the Twenty-First Century

In 1999, Turkey gained EU candidate status at the Helsinki Summit of the European Council, which triggered a major legal and institutional restructuring process in the early years of the twenty-first century (Kayasü and Yetişkul, 2014). That was also the year of the two consecutive devastating earthquakes in the country, the results of which were loss of life and property as well as of trust in the construction sector and the safety of urban areas. This was followed by the 2001 economic crisis. The IMF and World Bank programme in 2001 imposing new economic measures also affected the restructuring process in the country (Kuyucu, 2017; Çelik, 2020). Accordingly, several significant changes took place in the urban planning system and the institutional context of housing in the last two decades. Neoliberal urban policies consider urban land as a tool for economic development (Kayasü and Yetişkul, 2014). Thus, the macroeconomic policy adopted in this period by the government involved construction-led economic growth. This was supported by the deregulation and liberalization of the legal and institutional framework concerning urban development and its control (Balaban, 2008). In 2003, just after the Justice and Development Party (AKP) came to power, the government prepared an Urgent Action Plan (UAP). This plan's objective regarding "housing development and planned urbanization" defined two measures: (i) initiation of a countrywide housing programme to support owner-occupation among low-income households and (ii) prevention of squatter housing development through the redevelopment of squatter housing sites (Özdemir Sarı and Aksoy Khurami, 2018). In line with the UAP's objective, the Housing Development Administration (TOKİ), established in 1984, was restructured and engaged in housing production, mainly the production of subsidized owner-occupied flats (social housing programme of TOKİ) and the transformation of squatter housing areas. Before it was restructured, TOKİ was responsible mainly for providing credit support for construction. The primary beneficiaries of TOKİ were cooperatives. However, its role has changed to housing production, and the credit support provided to cooperatives stopped completely in 2005 (Türel and Koç, 2015). Throughout this restructuring period, TOKİ became a more powerful and autonomous institution regarding legal, administrative, and financial issues (Kayasü and Yetişkul, 2014). During 2003–2019, TOKİ produced nearly 845,000 housing units (TOKİ, 2019). Of this production, 86% is 'social-type housing projects', of which 17.4% is squatter housing renewal projects. Housing projects organized by TOKİ involve decisions external to local plans, and choices about the locations of these projects (on publicly owned land) have adverse effects such as urban spatial expansion and increased densities in inner-city areas.

In 2011, the Ministry of Environment and Urbanization (MEU) was established, replacing the Ministry of Public Works and Ministry of Environment and Forestry. The task of preparing legislation regarding settlement, development, and environment was given to the MEU. Furthermore, urban transformation implementations and their monitoring were also assigned to the MEU. In 2012, a new law concerning the transformation of areas at risk of disasters was enacted. This law is referred to as the 'urban transformation law' (Law no. 6306) and became a tool for transformation even though there was no risk of disaster (Uzun, 2019). In 2014, a new regulation concerning spatial plan preparation took effect in relation to the 1985 Development Law. These revisions to the 1985 Development Law are considered attempts at centralization (Özdemir Sönmez, 2017). In the current system, laws other than the Development Law define special types of plans like conservation plans, tourism plans, and industrial area plans, which are the relevant ministries' responsibilities (Özdemir, 2019).

In 2018, the presidential system replaced the parliamentary system in Turkey. This resulted in the transformation of legislative and institutional structures, the effects of which are still unclear. Furthermore, 2018 was the year of a new economic crisis. The effects of this crisis were also felt in 2019. Finally, the outbreak of Covid-19 in 2020 had negative impacts on construction and housing development. During this period of crisis, several types of government intervention have been seen. Tax incentives and mortgage interest cuts are some ordinary measures that attempted to increase housing demand. However, a new measure has to be emphasized here as no similar measures had ever been implemented previously in the Turkish Republic. This is a housing sales campaign led by the MEU. Throughout this campaign, the state provided support to some (selected) private construction firms to deplete their vacant housing stock that remained unsold. Today, despite the adverse effects of the 2018 crisis and the Covid-19 pandemic on housing production and transactions, it is a fact that there is a significant amount of surplus housing stock in the country. However, this housing output is not evenly distributed across the country (Özdemir Sarı, 2019).

The Supply Side: Housing Production, Existing Housing Stock, and House Prices

Housing production can be divided into two main groups of activity: construction of new dwellings and rehabilitation of existing housing stock. In line with this, for many countries, housing production statistics cover both new construction and rehabilitation. However, in Turkey, housing production means fundamentally the building of new houses. Rehabilitation and renovation decisions are comparatively rare and extremely difficult to follow from the current statistics. Thus, this part of the study focusses solely on new housing production. Turkey has a high housing production performance. However, the type of stock produced, production trends regarding dwelling sizes, and age distribution of the existing housing stock are also worth mentioning here.

Housing Production

Private sector investments have always dominated housing production in Turkey. In the pre-2000 period, in the absence of a comprehensive housing policy and social housing sector, state involvement in housing production as a direct actor was limited to cases in which a large number of dwelling units needed to be produced to house migrants and disaster victims (Tekeli, 1996). Since the 2002 elections, the public sector in Turkey has become a direct actor in housing production (Özdemir, 2011). After 2008, following the global financial crash and the slowing down of the Turkish economy, the role of large-scale infrastructure projects and housing projects in generating economic growth and employment has become more apparent (Kuyucu, 2017). In the last twenty years, sustained support for the construction sector and direct state involvement in housing production through TOKİ have resulted in increased housing output throughout the country. Figure 2.1 displays annual housing starts and housing completions during the period 1985–2020. As mentioned earlier, the increasing trend observed in completions in the last two decades reflects the recent legal changes that force construction firms and owners to apply for occupancy permits. However, housing starts (construction permits issued) are more realistic figures to evaluate the annual housing production in the country.

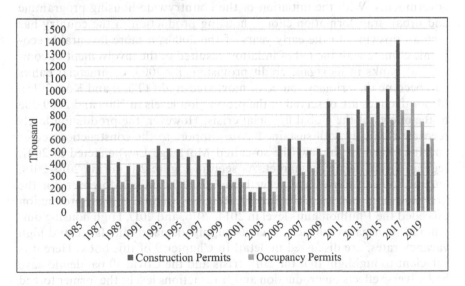

Figure 2.1 Annual housing starts and completions in Turkey during the period 1985–2020.

Source: Prepared by the author based on TURKSTAT (2014a, 2014b, 2021a).

The housing sector in Turkey was performing very well in new housing production even from 1985 to 1999, with an average of 438,000 annual housing starts. This period is characterized by high inflation rates and the 1994 economic crisis. The period ended with the August and November 1999 earthquakes, known as the Marmara earthquakes, which negatively impacted housing production levels. The primary factor contributing to the high performance displayed in Turkish housing production in the pre-2000 period was particularly the invention of 'flat ownership' relations, which gave rise to the construction of multi-unit structures in urban sites (Balamir, 1999). With the enactment of the Flat Ownership Law (Law no. 634) in 1965, the construction of blocks of flats became the dominant form of construction in the country. Amnesty Laws enacted in the second half of the 1980s also contributed to this production by transforming single-storey squatter houses into blocks of flats. As of 2019, 66% of all households in Turkey are estimated to live in flats (TURKSTAT, 2021b). However, the Covid-19 pandemic demonstrated that living in blocks is not tolerable when the time spent at home increases. During the lockdown periods, demand for single-family houses has increased, a trend that will be observed clearly in the permit statistics in the coming years.

The decline in the production levels during the period 2000–2002 can be attributed to the post-earthquake regulations that make it harder to construct buildings and the 2001 economic crisis. With the elections in 2002, the AKP came to power, and a new period began in the field of housing with gradually increasing control of land and housing markets by the AKP governments. With the initiation of the countrywide housing programme and urban transformation efforts, housing production in the country has seen an upswing. In the early years of the 2000s, a more favourable economic climate with the fall of inflation resulted in the involvement of commercial banks in mortgage credit provision. By 2004, commercial banks had become the primary source of mortgage credit (Türel and Koç, 2015). Minor declines are observed in the production levels in 2008 and 2009 due to the effects of the global financial crisis. However, the production levels recovered quickly with sustained state support to the construction sector and the effects of the law (the so-called Mortgage Law) enacted in 2007, regulating the mortgage market. As seen in Figure 2.1, during 2010–2018, annual housing starts were never below 600,000 units. Moreover, for the first time in the history of the Turkish Republic, annual housing production exceeded the 1 million units level in 2014, 2016, and 2017. High housing output and its negative impacts, such as supply–demand imbalance and high vacancy rates, are discussed in detail in Chapter 9 of this book. Here it is sufficient to highlight that the 2018 crisis and the Covid-19 pandemic have had adverse effects on production and transactions levels; the former forced the government to launch a housing sales campaign to deplete the excess stock in the hands of private construction firms.

Table 2.1 Housing production by type of investor (dwellings in thousands): 2002–2020

Years	Private sector	Public sector	Cooperatives
2002	131	7	24
2004	284	18	27
2006	519	28	51
2008	410	59	32
2010	770	80	52
2012	683	73	11
2014	961	51	16
2016	922	67	13
2018	607	50	9
2020	506	40	8

Source: Prepared by the author based on TURKSTAT (2021a).

As mentioned above, the last two decades have witnessed the state's increasing involvement in housing production. Table 2.1 displays housing production levels by type of investors during the period 2002–2020. Accordingly, annual housing production by the public sector has increased significantly in the last two decades. However, as of 2020, 91% of all dwelling units produced are through private sector investments (TURKSTAT, 2021a). At the same time, cooperatives' contribution to total production has visibly decreased and almost come to an end, particularly in the last decade.

As Table 2.1 displays, private sector investments in housing production are still the dominant form of investment in the housing sector. That is also the reason for the uneven distribution of the housing output throughout the country. As might be expected, private companies' investment decisions and location choices are determined particularly by profit expectation. The lack of policy measures to channel the investments according to the requirements in different regions of the country results in an oversupply of housing in some regions and housing shortages in others.

Compared to most European countries, the size of the private sector engaged in building construction in Turkey is significantly large in terms of companies and employment. As of 2019, based on Company Accounts Statistics (TURKSTAT—CBRT, 2019), there are nearly 83,000 companies in Turkey engaged in building construction. Enterprises engaged thus constituted almost 11.4% of the total firms and recruited 6% of the total employees in the country in 2019. Many shutdowns were observed among the construction firms with the 2018 economic crisis. The crisis led to more than half of the medium-sized and large enterprises being closed (TURKSTAT—CBRT, 2019). The effects of the Covid-19 pandemic on the size of the construction sector, however, currently remain unclear. Building construction companies constitute 72% of the construction sector enterprises and 60% of the employees in the sector.

Physical Features of the Existing Housing Stock

Data regarding the physical features of the existing housing stock are difficult to find in official statistics. However, representative sample surveys such as the Household Budget Survey (HBS) and Survey of Income and Living Conditions (SILC) conducted by TURKSTAT present the overall features of the existing stock in the country. According to the HBS, in line with the discussion presented above about the housing production performance of the country, it is not surprising to observe that the Turkish housing stock is relatively new, with nearly 55% of it built in the last three decades (Figure 2.2). As Figure 2.2 displays, dwellings older than 60 years make up only a 5% share of the total stock. This is due to the fact that throughout the history of the Turkish Republic demolishing and rebuilding are seen as an easy and profitable way under the conditions of increased development rights. Current urban transformation policies also follow this conventional trend, leaving no room for rehabilitation investments. However, it is clear that development rights and densities cannot be increased limitlessly. Thus, rehabilitation policies are inevitable necessities for Turkey in the coming decades, not only for ensuring efficiency of use of the existing stock but also for assuring safety in built environments (Özdemir Sarı, 2019).

Another significant piece of information that can be derived from Figure 2.2 is a continuous trend for increasing dwelling sizes in new housing production in the country. As of 2019, 3–4-bedroom dwelling units dominate the existing housing stock, making up 85% of the total stock (TURKSTAT, 2021b).

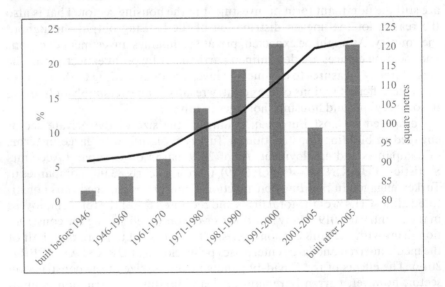

Figure 2.2 Age distribution of dwellings and mean dwelling sizes in Turkey as of 2018.

Source: Prepared by the author based on the raw data of the Household Budget Survey-2018 compiled by TURKSTAT.

However, the demographic trends discussed below indicate that one- and two-person households have been rising in the last two decades. Whether these conflicting trends reflect a supply–demand imbalance or increased demand for larger dwellings requires more detailed investigation.

House Prices

In Turkey, there is no reliable information to enable house price development to be examined. The house price index has been prepared by the Central Bank of the Republic of Turkey (CBRT) since 2010. It is based on expert evaluation reports prepared for dwellings subject to valuation due to mortgage applications and thus underrepresents the Turkish housing market (Özdemir Sarı, 2019). According to house sale statistics (TURKSTAT, 2021c), the highest rate of mortgage sales is observed for 2020, 38% of all sales. This rate was due to the interest rate cuts implemented as a measure to increase transactions in 2020. This indicates that mortgage finance is still not an option for the wider segments of society. Therefore, the house price index does not represent the overall Turkish housing market. Nevertheless, Chapter 6 of this book discusses the house price index and the changes in it comprehensively for the country overall and major cities. It is sufficient to say here that following the production boom of 2014–2017 and with the increasing trend of inflation rates, real prices started to decline and even negative price growth was observed starting in the third quarter of 2017. This negative trend in prices deepened during the economic crisis in 2018–2019. Covid-19 lockdowns in 2020 increased the demand for residential mobility, favouring less dense urban areas, single-family houses, and coastal settlements. Combined with the mortgage rate cuts in 2020 and low housing production levels during 2018–2020, positive growth was observed in real house prices in 2020.

The Demand Side: Demographic Trends, Tenure Structure, Housing Wealth, and Affordability

Changes in the social and demographic structure have also been observed over the last two decades. Demand-side changes such as population trends, household formation rates, and household composition are significant in the context of housing since they are expected to be responded to by the supply side in terms of the number and size of the new houses constructed. However, as discussed above, housing production in Turkey appears to be unresponsive to changes in demographic trends. This results in supply–demand mismatch, leading to high vacancy rates on the supply side and affordability problems on the demand side. In order to highlight demand-side dynamics, this part of the study focusses on demographic trends that trigger demand for housing, changes in tenure structure and rise of a new tenure category, and distribution of housing wealth and affordability problems in the society.

Demographic Trends and Demand for Housing

The early decades of the Turkish Republic were characterized by rapid population increase and rapid urbanization. The population of the country tripled in almost five decades between the 1920s and 1970s. In the last 50 years, this trend in population increase has continued though at a comparatively slow pace. The country's population has increased almost 2.4-fold from 35.6 million in 1970 to 83.6 million in 2020 (TURKSTAT, 2021d). This increased population, in turn, stimulated housing demand. In the same period, the number of households increased 3.9-fold from 6 million to 24 million, which has further driven the demand for housing (TURKSTAT, 2015, 2021e). Figure 2.3 demonstrates these demographic trends. Furthermore, during 2010–2020, the population increased by almost 9.9 million and over 35% of this increase was accounted for by Syrian refugees. As of 2018, only 6.4% of the Syrian refugees live in camps, while 93.5% prefer to live outside the camps, particularly in metropolitan cities (Güngördü and Bayırbağ, 2019). This has also accelerated housing demand in the country.

Moreover, the household structure is also influential in increasing housing demand. The average household size in Turkey has followed a declining trend; it was 5.6 in 1970 and had fallen to 4.5 by 2000. It has continued to decrease in the last two decades from 4.0 in 2008 to 3.3 in 2020. This consistent fall in household size and the increasing number of households show that the rate of small households has increased in the society, whereas

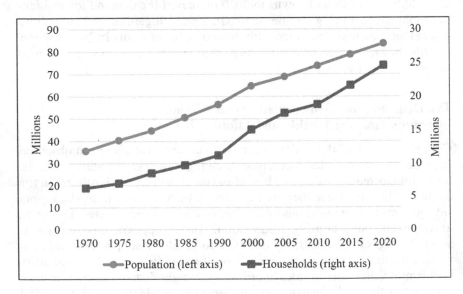

Figure 2.3 Demographic trends during the period 1970–2020.

Source: Prepared by the author based on TURKSTAT (2015, 2021b, 2021d, 2021e).

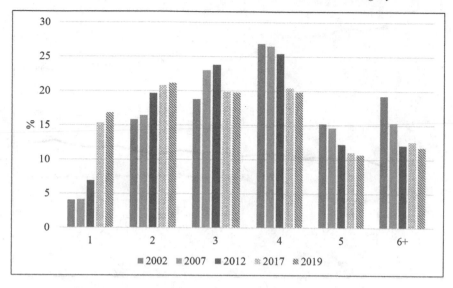

Figure 2.4 Changing household size structure during 2002–2019.

Source: Prepared by the author based on TURKSTAT (2021b).

the rate of extended families has declined. Figure 2.4 displays the changing household size structure in the first two decades of the 2000s. Accordingly, during 2002–2019, a dramatic change was observed in one-person households, increasing from 4% to 17%. This shift in the household size structure is a result of the combination of economic, social, and cultural factors. More young people currently leave their parental home for university education, to get a job, or due to work-related mobility. Women are more involved in the workforce and the marriage age has increased. Furthermore, divorce rates have increased. In addition, as life expectancy has increased, the share of widows living alone has increased. At the same time, for households with four or more people, a significant decline is observed. Family ties are still strong in society, but living in extended families is preferred less. All these demographic factors have significant effects on housing demand. It should be noted that apart from the demographic factors, a comparatively good economic climate with economic growth and rising per capita GDP levels for most of the years in the last two decades have also been influential in increasing housing demand.

Changes in Housing Tenure Structure

As mentioned above, the mortgage finance system in Turkey was developed in 2007. Before that time, housing finance was largely dependent upon state-owned institutions; yet a very small portion of the population benefited

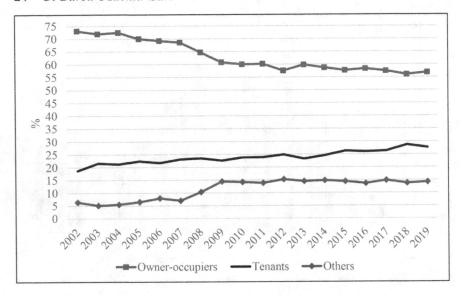

Figure 2.5 Changing tenure structure during 2002–2019.

Source: Prepared by the author based on TURKSTAT (2021b).

from the credit support provided. Due to the shortage of capital resources with respect to demand and high inflation levels, commercial banks did not provide any finance for housing until the 1990s (Türel, 1994). Even under those conditions, homeownership has been the dominant tenure type in the country (Sarıoğlu-Erdoğdu, 2010). In such a system, households' savings serve as the primary source of capital for becoming a homeowner (Özdemir Sarı and Aksoy Khurami, 2018). Figure 2.5 displays the changing tenure structure during the period 2002–2019.

It should be noted that Figure 2.5 represents the tenure structure for Turkey overall as it has not been possible to differentiate rural/urban areas from the official statistics since 2013. This means tenancy, a particularly urban phenomenon, is underrepresented in the figure, whereas owner-occupancy is overrepresented. As Figure 2.5 displays, the proportion of owner-occupiers declined from 73% to 57% during 2002–2019. Some part of this fall has been revealed as increasing tenancy, whereas the rest displays an emerging tenure type labelled 'others' in the figure. The 'others' category here represents households who are neither owner-occupiers nor tenants. More clearly, these households occupy their parents'/relatives' houses at below-market rent levels or pay no rent at all. This category also represents the increased multiple ownership in the country in the last two decades. In other words, housing policy aiming to encourage owner-occupation among low-income households has increased the multiple ownership in the society for those who are already homeowners and have additional means to buy extra houses.

Figure 2.5 also reveals that the tenure structure has changed in favour of tenants in the last two decades, displaying a 20–30% portion among all tenure modes. However, compared to the pre-2000 period, when tenancy in urban areas had a share of 35–40% among the tenure modes, these figures are comparatively low. There is no policy in Turkey to keep the size of the rental sector at a determined scale. Instead, the rented stock continues to grow or decline as a result of the policies supporting homeownership and trends in society.

Housing Wealth and Affordability

The policies encouraging owner-occupation and supporting new housing production since the early 2000s had some wealth and affordability effects that are not monitored by related government agencies. Although limited, research on housing wealth and affordability in Turkey is emerging (see, for instance, Aksoy, 2017; Aksoy Khurami 2021). An in-depth investigation of wealth and affordability is not within the scope of the present study. However, it could be useful to provide some hints about the current distribution of housing wealth and affordability problems in society. For this purpose, cross-sectional data of HBS and SILC 2018 are employed, and the relevant information is presented in Table 2.2.

According to Table 2.2, comparatively high-income households obtain greater shares from total disposable income and are less cost-burdened by housing expenditures. Households in the lowest quintile have only 6.6% of the total disposable income and devote 35% of their budget to housing expenditures on average. Of these households, 20.4% report experiencing high housing cost burden. However, not only the households in the lowest-income quintiles but also those in the low- and medium-income quintiles devote a significant share of their budget to housing expenditures. Although the owner-occupancy rate is quite similar among different income categories, probably a result of the homeownership policies implemented in the last two decades, this appears not to be a solution for housing affordability problems experienced at the lower end of the income scale. The last three columns of Table 2.2 are considered to reflect the housing wealth of households. It is seen that dwellings occupied by households in the highest income quintile have 3.8 times higher mean value compared to those occupied by households at the lowest end of the income scale. In other words, being an owner-occupier does not mean having the same level of wealth as all owner-occupiers. Furthermore, multiple ownership is much greater at the highest end of the income scale. The share received from the rental revenue obtained through multiple ownership is significantly high for households having the highest income. Considering that rental property income is included in total disposable income, one can conclude that housing wealth, particularly wealth derived from multiple ownership, significantly contributes to society's income inequality.

Table 2.2 Housing wealth and affordability by income quintiles

Income quintile	Income quintile's share of the total equivalised income (%)	% of households cost-burdened by housing expenditure	Share of housing expenditure in household budget (%)	% owner-occupier	Mean house values of owner-occupiers (thousands, TRY)	% of multiple homeownership	Income quintile's share of the rental income received from multiple ownership (%)
Lowest	6.6	20.4	35	60.5	88.6	3	2
Low	11.2	13.4	28.1	60	123.4	5.8	5.5
Medium	15.2	11	24.2	61.4	146.5	7.8	10.6
High	21.4	7.2	20.8	61.9	177.6	9.5	19
Highest	45.6	4	13.8	59.9	332.9	15.9	63

Source: Prepared by the author based on the raw data of the Household Budget Survey-2018 and Survey of Income and Living Conditions-2018 compiled by TURKSTAT.

Conclusions

In this chapter, an overview of the Turkish housing system is presented to provide the essential background information necessary for the rest of the chapters in the book. The chapter is organized into three basic sections reflecting institutional context and supply and demand components of the housing system. The last two decades have witnessed significant changes in the institutional context of housing development. Currently, control over land and housing markets in Turkey is highly centralized. A significant amount of housing output is created with the help of government support and direct involvement in production in line with the macroeconomic policy dependent on construction-led growth. Despite the increasing role of the state in housing production, private investments still dominate the market. The dominant form of construction is blocks of flats and there is an apparent trend towards building larger homes. On the demand side, increasing numbers of one- and two-person households and declining average household size indicate that the society's social and demographic structure is changing. However, the production side is highly unresponsive to these trends. Continuous support for homeownership and housing production has had some results in terms of housing wealth. Although owner-occupation has been achieved by almost all segments of the society at similar rates, wealth effects are basically secured for the highest income households through increased multiple ownership and rental property income. This contributes further to the inequalities already existing in society. Affordability, on the other hand, is still a relevant issue for Turkish households.

References

Aksoy E (2017) Housing affordability of different income groups in Turkey: Regional comparison. Ankara: Middle East Technical University. Available at: http://etd.lib. metu.edu.tr/upload/12620805/index.pdf. Accessed 1 July 2021.

Aksoy Khurami E (2021) Failing promises of homeownership in Turkey. Ankara: Middle East Technical University. Available at: https://tez.yok.gov.tr/UlusalTezMerkezi/ TezGoster?key=wf-FPgY-5qjHEzEoOgvMsxU5qCRrYLKcFwej-DLxoo PO7Rv2QCNVLHuXiQJcCko8. Accessed 1 July 2021.

Balaban O (2008) Capital accumulation, the state and the production of built environment: The case of Turkey. Ankara. Middle East Technical University. Available at: http://etd.lib.metu.edu.tr/upload/12609418/index.pdf. Accessed 15 June 2021.

Balamir M (1996) Making cities of apartment blocks: Transformation of the built environment in Turkey by means of reorganizations in property rights. In: Sey Y (ed) *Housing and Settlement in Anatolia: A Historical Perspective.* İstanbul: Türkiye Ekonomik ve Toplumsal Tarih Vakfı, 335–344.

Balamir M (1999) Formation of private rental stock in Turkey. *Netherlands Journal of Housing and the Built Environment*, 14(4), 385–402.

Çelik Ö (2020) The AKP's income-differentiated housing strategies under the pressure of resistance and debt. In: Bedirhanoğlu P, Dölek Ç, Hülagü F and Kaygusuz Ö (eds) *Turkey's New State in The Making: Transformation is Legality, Economy and Coercion.* London: Zed, 151–164.

Duyguluer F (2006) İmar mevzuatının kayıpları (Losses of the development legislation). *Planlama*, 2006(4), 27–37.

Ersoy M (2011) Some observations and recommendations on the practice of upper level urban plans in Turkey in the light of sustainable development. Tokyo: APSA Congress. Available at: http://www.melihersoy.com/wp-content/uploads/2012/04/Some-Observations-and-Recommendations-on-the-Practice-of-Upper-Level-Urban-Plans-in-Turkey-B.101.pdf. Accessed 30 June 2021.

Güngördü FN, Bayırbağ MK (2019) Policy and planning in the age of mobilities: Refugees and urban planning in Turkey. In: Özdemir Sarı ÖB, Özdemir SS, Uzun N (eds) *Urban and Regional Planning in Turkey*. Switzerland: Springer, 185–211.

Gyourko J, Molloy R (2015) Regulation and housing supply. In: *Handbook of Regional and Urban Economics 5B*. Amsterdam: North-Holland, 1289–1337.

Harsman B, Quigley JM (1991) Housing markets and housing institutions in a comparative context. In: Harsman B and Quigley JM (eds) *Housing Markets and Housing Institutions: An International Comparison*. Springer Science & Business Media, 1–29.

Kayasü S, Yetişkul E (2014) Evolving legal and institutional frameworks of neoliberal urban policies in Turkey. *METU JFA*, 31(2), 209–222.

Keleş R (2002) *Kentleşme politikası (Urbanisation Policy)*. 7th Edition. Ankara: İmge.

Kuyucu T (2017) Two crises, two trajectories: The impact of the 2001 and 2008 economic crises on urban governance in Turkey. In: Adaman F, Akbulut B and Arsel M (eds) *Neoliberal Turkey and Its Discontents: Economic Policy and the Environment Under Erdogan*. London: I.B. Tauris, 44–74.

Oxley M, Haffner H (2012) Comparative housing research. In: Smith SJ, Elsinga M and O'Mahony LF et al. (eds) *International Encyclopedia of Housing and Home*. Amsterdam: Elsevier, 199–209.

Ozkan HA, Turk SS (2016) Emergence, formation and outcomes of flexibility in Turkish planning practice. *IDPR*, 38(1), 25–53.

Özdemir D (2011) The role of the public sector in the provision of housing supply in Turkey, 1950–2009. *International Journal of Urban and Regional Research*, 35, 1099–1117.

Özdemir SS (2019) A new route for regional planning in Turkey recent developments. In: Özdemir Sarı ÖB, Özdemir SS, Uzun N (eds) *Urban and Regional Planning in Turkey*. Switzerland: Springer, 13–37.

Özdemir Sarı ÖB (2019) Redefining the housing challenges in Turkey: An urban planning perspective. In: Özdemir Sarı ÖB, Özdemir SS, Uzun N (eds) *Urban and Regional Planning in Turkey*. Switzerland: Springer, 167–184.

Özdemir Sarı ÖB, Aksoy Khurami E (2018) Housing affordability trends and challenges in the Turkish case. *Journal of Housing and the Built Environment*. https://doi.org/10.1007/s10901-018-9617-2

Özdemir Sarı ÖB, Özdemir SS, Uzun N (2019) Evaluation of the issues and challenges in Turkey's urban planning system. In: Özdemir Sarı ÖB, Özdemir SS, Uzun N (eds) *Urban and Regional Planning in Turkey*. Switzerland: Springer, 281–287.

Özdemir Sönmez N (2017) Planlamada yeni yasal düzenlemeler (New legal arrangements in planning). In: Özdemir SS, Özdemir Sarı ÖB, Uzun N (eds) *Kent Planlama*. Ankara: İmge, 643–665.

Sarıoğlu-Erdoğdu GP (2010) A comparative analysis of entry to home ownership profiles: Turkey and the Netherlands. *METU JFA*, 27(2), 95–124.

Schill MH (2005) Regulations and housing development: What we know. *Cityscape*, 8(1): 5–19.

Tekeli İ (1996) *Türkiye'de yaşamda ve yazında konut sorununun gelişimi (The development of the housing problem in Turkey in life and in literature)*. Ankara: ODTÜ Basım İşliği.

TOKİ (2019) TOKİ corporate profile document. Available at: http://i.toki.gov.tr/content/entities/main-page-slider/20191011095737969524-pdf.pdf. Accessed 20 June 2021.

TURKSTAT (2014a) Construction permits: New buildings and additions by use of building (old classification): 1954–2003. Available at: https://data.tuik.gov.tr/Kategori/GetKategori?p=insaat-ve-konut-116&dil=1. Accessed 20 June 2021.

TURKSTAT (2014b) Occupancy permits: Completed or partially completed new buildings and additions by use of building (old classification): 1964–2003. Available at: https://data.tuik.gov.tr/Kategori/GetKategori?p=insaat-ve-konut-116&dil=1. Accessed 20 June 2021.

TURKSTAT (2015) Total household population, number of households and average size of household. Available at: https://data.tuik.gov.tr/Kategori/GetKategori?p=nufus-ve-demografi-109&dil=1. Accessed 25 June 2021.

TURKSTAT (2021a) Building permits statistics (2002–2021). Available at: https://biruni.tuik.gov.tr/yapiizin/giris.zul?dil=ing. Accessed 20 June 2021.

TURKSTAT (2021b) Household consumption expenditure statistics (2002–2019). Available at: https://data.tuik.gov.tr/Kategori/GetKategori?p=Income,-Living,-Consumption-and-Poverty-107. Accessed 1 June 2021.

TURKSTAT (2021c) House sales in detail of mortgaged and other houses by provinces and years, 2013–2021. Available at: https://data.tuik.gov.tr/Kategori/GetKategori?p=Insaat-ve-Konut-116. Accessed 1 July 2021.

TURKSTAT (2021d) Population of province/district centers and towns/villages by years and sex, 1927–2020 Census of Population—ABPRS. Available at: https://data.tuik.gov.tr/Kategori/GetKategori?p=nufus-ve-demografi-109&dil=1. Accessed 30 June 2021.

TURKSTAT (2021e) Number of households by size and type. Available at: https://data.tuik.gov.tr/Kategori/GetKategori?p=nufus-ve-demografi-109&dil=1. Accessed 25 June 2021.

TURKSTAT-CBRT (2019) Company accounts statistics (2009–2019). Available at: https://data.tuik.gov.tr/Bulten/Index?p=Sektor-Bilancolari-2019-33602. Accessed 1 July 2021.

Türel A (1994) Housing finance in Turkey during the last decade. In: Bartlett W, Bramley G (eds) *European Housing Finance: Single Market or Mosaic?* Bristol: SAUS Publications.

Türel A, Koç H (2015) Housing production under less-regulated market conditions in Turkey. *Journal of Housing and the Built Environment*, 30(1), 53–68.

Uzun N (2019) Transformation in residential areas: Regeneration or redevelopment? In: Özdemir Sarı ÖB, Özdemir SS, Uzun N (eds) *Urban and Regional Planning in Turkey*. Switzerland: Springer, 151–166.

Van der Heijden H (2013) *West European Housing Systems in a Comparative Perspective*. Amsterdam: IOS Press.

3 Housing Policies in Turkey

Paving the Way for Urban Transformation

Nil Uzun

Introduction

Housing has always been a significant issue for policymakers in Turkey and housing policies have changed in parallel with urbanization and economic policies. Four basic periods can be determined by taking several milestones related to housing policies into consideration. During the first period, starting in 1923, housing provision for government officials was one of the main concerns following the recovery from the two world wars. In addition, inadequate housing supply resulting from rapid urbanization necessitated new policies to cope with this housing crisis towards the end of the 1950s. The 1961 Constitution and the First Five-Year Development Plan (FYDP) announced in 1962 are two milestones denoting the beginning of the second period. During this period, which lasted until the 1980s, 'providing housing for all' was the main discourse in housing policies. Economic policies aiming at stability marked the beginning of the third period. A milestone in this period was the Mass Housing Law (Law no. 2487) enacted in 1981. Its purpose was to solve the problems in the housing market resulting from the impact of economic policies implemented during the previous period. In this period, the effects of the law and mass housing production were observed widely. During the fourth period, starting in 2002, political changes had substantial implications for housing policies, and regulations concerning urban transformation came to the fore. In the following sections, the evolution of housing policies in Turkey is explored with reference to these four periods.

Housing Provision for Government Officials and the Housing Crisis (1923–1960s)

The period between 1923 and the 1960s was characterized by restructuring and foundation. After the foundation of the Turkish Republic, between 1923 and 1929, the aim was to achieve economic independence and rapid economic development. The developments experienced caused protectionist and statist policies to rise to prominence and formed the basis for the beginning of a new era in this sense. While a more intense mixed economy system

DOI: 10.4324/9781003173670-4

compared to the first years of the republic was adopted from 1930 to 1939, serious breakthroughs were made in industrialization as part of the First Five-Year Industrial Plan. Although Turkey followed an impartial policy in the 1940s and stayed out of the Second World War, its economy could not avoid the devastating effects of the war. Controls over the private sector increased while statist policy practices remained in existence. Foreign aid was accepted for the first time during this period. Following the election of a new ruling party in 1950, the development strategy changed radically. The intensely statist viewpoint was abandoned and the promotion of private enterprise gained prominence (Çoban, 2012).

As well as economic restructuring, the young Republic was faced with a massive development and housing problem. The reconstruction of Western Anatolian cities, as they had been ravaged by the Independence War, and the resettlement of the Turkish population arriving in the country as a result of the population exchange required the construction of many houses. There was also the problem of the restructuring and foundation of Ankara as a new and modern city after it was declared the capital and, therefore, providing housing for government officials. Private enterprises were encouraged to undertake the construction work to provide the new housing stock. Moreover, activities to support social development, such as the modernization of housing in rural areas, were also on the agenda. In addition, after the Erzincan earthquake in 1939, there were attempts to produce earthquake-resistant housing (Sey, 1998; Tekeli, 2012).

In the 1924 Constitution, it was stated that the house is protected against all kinds of attacks. In the Constitution, the right to privacy of housing was also accepted. However, apart from these points related to individual rights, no decree regulating the right to housing was found. As the economy was predominantly based on agriculture, most of the population was still living in rural areas. In 1927, 84% of the total population was living in rural areas and this rate had decreased only to 81% by 1950 (İçduygu and Sirkeci 1999). Therefore, the presence of a stagnant urban population was not a compelling factor in the development of a housing policy and the housing problem in rural areas was not regarded as a priority by policymakers. The scarce resources were allocated for industrialization rather than housing, in line with the statist policies of this period, laying the foundations for industrial capitalism. Therefore, it is not possible to talk about a comprehensive housing policy for this period (Çoban, 2012).

On the other hand, there were new regulations to help meet the housing needs of government employees, whose numbers increased with the development of the institutional organization of the new state. A law enacted in 1925 (Law no. 586) allowed government employees to be given an advance of half their salaries to establish housing cooperatives. In another law enacted in 1928 (Law no. 1352), the Ministry of Finance was authorized to build housing for government employees with help from the Treasury. However, this law did not result in any building taking place. In 1929, due to another new law (Law no. 1452), housing compensation started to be paid to government employees and this continued until 1951 (Keleş, 2013).

Furthermore, the Municipality Law enacted in 1930 (Law no. 1580) gave the task of constructing cheap municipal houses and selling them to municipalities. However, the municipalities did not undertake any housing works since they were unable to allocate a budget for this optional task without neglecting their mandatory duties. Later a decree was added to the Municipality Law, authorizing the municipal councils to determine the task of constructing municipal residences and leasing or selling them as their mandatory duties. Nevertheless, municipalities have not made an effort to intervene in the housing problem using this authority. On the other hand, in the 1930s, special attention was paid to the construction of workers' and peasants' houses, and the government programme envisaged that municipalities would build houses. While the practices in parallel with the policies followed in the 1930s had just begun, a new period of economic depression and stagnation for the country was triggered by the Second World War and, after 1939, housing production gradually decreased (Sey, 1998; Çoban, 2012).

In 1944 a new law (Law no. 4626) was passed to deal with the issue of building housing for government employees. It authorized the Ministry of Development to build housing for these citizens where needed. However, Ankara, the capital, was given priority. The Saraçoğlu neighbourhood in Ankara consists of 450 residences built with authority granted by this law. In this neighbourhood, mostly senior officials of the Ministries of National Defence, Interior, Finance, and Justice lived as tenants (Keleş, 2013). Another problem that was focussed on by the state in the 1940s was workers' housing. Priority was given to the construction of housing for the workers and officials of the newly established factories in Anatolia. Although these dwellings were inadequate, it was an important initiative for that time (Sey, 1998).

Along with new regulations, several institutions providing loans for housing were founded during this period. In 1926 the state-owned Real Estate and Orphans Bank, later renamed the Real Estate and Credit Bank of Turkey in 1946, was established in order to support construction initiatives, protect the rights of orphans, and offer loans and credit required for housing. Social security institutions also contributed to housing production by providing housing loans to their members. The Workers' Insurance Institution, which was renamed the Social Insurance Institution (SSK) in 1965, started issuing housing loans in 1950. Unlike the Real Estate and Credit Bank of Turkey, this institution gave housing loans only to members of housing cooperatives. The SSK contributed to the production of more than 200,000 houses over three decades with such loans. The State Railways Workers Pension Fund and the Military Factories Retirement and Assistance Fund, which operated until the SSK was established, also provided housing loans to its members (Çoban, 2012).

The provision of housing through cooperatives became another solution regarding housing supply as rents increased and a housing shortage occurred. The first housing cooperative, namely the Bahçelievler Housing Cooperative, was established in 1934 in Ankara. The Real Estate and Credit Bank of Turkey provided credit to meet about 90% of the construction costs of houses of the Bahçelievler Housing Cooperative. Members of the cooperative were high-level bureaucrats and they were attempting to use their positions to solve their housing problems. Their aim was not only to have somewhere to live but also to benefit economically from the cooperative, which was also regarded as a profit-making means. The houses constructed were turned over to the partners of the Bahçelievler Housing Cooperative in 1938. This number was greater than the sum of the building permits issued by the Ankara Municipality in those years. Consequently, due to economic crises and lack of resources, cooperatives were regarded as an alternative to state and private sector housing provision (Sey, 1998; Özkan, 2009; Tekeli, 2012).

During this time, along with the implementations favouring owner-occupation, in order to protect tenants, housing rents were pegged at the 1939 values by the National Conservation Law of 1940 (Law no. 3780). Although it can be argued that the powers such as price and rent control granted to the public administration by this law aimed to ensure that the society would be affected less by the negative economic effects of the Second World War, the implementation of the National Protection Law, on the whole, resulted in either protection or strengthening of large capital. In 1947 a 20% increase in rents and talk of abandoning the rent limitation due to the end of the war came to the fore. However, with another law enacted in 1955 (Law no. 6570), rents were once more pegged at the 1939 values and this practice continued until the 1960s (Çoban, 2012).

Migration from rural to urban areas started to accelerate in the 1950s. This movement was driven by factors such as the mechanization and modernization of agriculture, change in the traditional land ownership regime, land deprivation or consolidation of lands, and the development of transportation facilities (İçduygu and Sirkeci, 1999). Lack of an adequate housing supply combined with the rapid migration of the 1950s and 1960s resulted in squatter housing development in larger cities like Ankara, İstanbul, and İzmir. In 1948 the first law was enacted (Law no. 5218) to improve the condition of squatter houses within the borders of Ankara Municipality and to prevent squatter houses by providing ten-year interest-free loans, for those building new houses, to buy land. Land owners were obliged to construct a building within a maximum of two years. The same year, a second law (Law no. 5228) was enacted to enable those who received land assistance via the previous law (Law no. 5218) to benefit from housing loans with 5% interest up to 75% of the construction price. Initially, it only applied to land in Ankara, but later its scope was expanded to other cities. Taking advantage of these two laws, Ankara Municipality established the neighbourhood

Yenimahalle, which was designated as a new settlement area other than squatter housing. The first constructions started in 1949 on land prepared by the Municipality, the development and infrastructure of which were complete, and the district envisaged at the beginning was completed within five to six years. Although the aforementioned two laws were envisaged mostly to enable low-income people to own houses, the example of Yenimahalle ended up benefitting the middle class. Adopting this way of selling cheaply produced public land had negative consequences also in terms of housing policy such as not being able to prevent the privately owned land and house from causing speculative gains during the subsequent development of the cities. Regarding the complaints about the acceleration of squatting and damage to property rights, a new law envisaging the prevention of squatting and the demolition of squatter houses was enacted in 1949 (Law no. 5431) (Çoban, 2012; Genç, 2014).

In 1953, the Law on Encouragement of Building Construction (Law no. 6188) abolished the laws numbered 5218 and 5228 and aimed to prevent squatting by introducing similar provisions and to increase the production of housing by selling cheaply produced public land. It, in turn, was replaced by another law on squatter houses in 1966. However, neither goal was achieved. In 1958 the Ministry of Development and Settlement was established (Law no. 7116). One of the duties of the Ministry was defined as determining the principles of housing policy suitable for the country's structure and ensuring their implementation. The necessity of solving the squatter housing problem via a policy implemented throughout the country was also mentioned in the law (Çoban, 2012; Genç, 2014).

During these years, as the impact of the housing problem increased, new actors started to enter the housing market. Until the mid-1950s, the rule of building a single house on a single building plot was in force. In this case, the owner of the building plot had a house built on it, mostly for his or her own use. This made it difficult for the entrepreneur who wanted to build a house to sell it by considering the exchange value of the house to enter the housing market. With an amendment made to the Land Registry Law (Law no. 2644) in 1954, this rule was changed and the establishment of the easement was provided to benefit from one floor or flat of a building to be built on a building plot. Thus, when more than one person was allowed to establish an easement on a building constructed on a single building plot, the phenomenon of building and selling emerged[1] (Çoban, 2012).

Five-Year Development Plans: Providing Housing for All (1960s–1980s)

Following the political instability of the 1950s, the military coup of 1960, and the ongoing economic depression, a period of planned industrialization started in Turkey. The 1961 Constitution and the First FYDP, announced in 1962, are two milestones marking the beginning of this period, which can also be referred to as the planned development period. The import-substituting industrialization strategy was explicitly implemented as an economic policy. In this context, while ensuring that import substitution policies were executed under the guidance of development plans, the focus was on the domestic production of non-durable consumer goods and industrial products.

The housing problem faced by low-income people, which could not be resolved with various laws and financial systems, was the subject of a constitutional regulation for the first time in 1961 in which the right to housing was included. However, it was regulated under the title of 'right to health'. It was stated that the need for housing should be met as a condition of living a life in physical and mental health. The state was responsible for taking the necessary measures to meet the housing needs of poor and low-income families in accordance with their health conditions. As the scope of poverty widened due to factors such as the general economic depression and the course of the world economy, that is, as middle-income families gradually faced difficulties making ends meet, the scope of the duty of the state in this area also expanded (Çoban, 2012).

In 1965, the Flat Ownership Law (Law no 634) was enacted in order to solve the housing crisis. The ownership pattern in a block of flats with many shareholders was determined on the basis of their share of the land before this law came into effect. With this law having the independent ownership right for a separate unit of a building or a share in a block of flats before it was built became possible and erecting a multi-storey block of flats on one building plot with a small amount of capital turned out to be easier. Therefore, the legislation led to an increase in the number of flats in buildings. The allocation of the flats was determined after the completion of the construction. This new ownership system became a powerful mechanism to obtain effective capital for major investments. As a result, many new, high-density neighbourhoods were created, whereas some low-density districts were even demolished and rebuilt with higher density (Balamir, 1999).

During this period, a comprehensive housing policy was maintained in the light of the FYDPs envisaged with the establishment of the State Planning Organization (DPT) in 1960. During the First FYDP (1963–1967) period, housing problems were handled as a whole for the first time and their relationship with development was established. In the plan, housing investments were considered economically unproductive. Therefore, keeping housing sector investments below 20% of the total investments was one of the main principles. The aim was to build more houses within these limits. The way to do this was to reduce the surface area of the houses to be built. For this reason, public loans were given to houses not exceeding 69 square metres at the beginning and 100 square metres later, and small houses were encouraged. As a means of reducing the cost of housing, not only decreasing the areas, but also the prevention of land speculation, the taxation of land value increases during purchases and sales, and the support of mass housing institutions such as cooperatives were among the recommendations of the First FYDP. Considering squatter houses not as illegal structures but as a social and economic phenomenon for the first time was a change in attitude brought about by this plan (Keleş, 2013). The Squatter Law enacted in 1966 (Law no. 775) was an expression of this change. As in previous laws, improvement, clearance, and prevention of squatter houses were the main aims. It was decreed that municipalities should acquire land for the purpose of preventing squatter houses and use these plots in order to eventually leave them entirely to the citizens. This law envisaged various aids to meet the housing needs of those who received land aid. One of them was to provide financial aid from squatter house funds. These funds were created under the supervision of the municipalities and the Ministry of Development and Settlement (Genç, 2014). The First FYDP also proposed to encourage the construction of low-rent public housing as well as to reduce the pressure of high housing rents on low-income people. For this purpose, a rent policy that would protect tenants and landlords in accordance with the objectives of the housing policy was employed. Moreover, it was aimed to eliminate the irregularity in urbanization and the imbalance in the settlement order by stipulating that urbanization should be proportional to the job opportunities created in the cities (DPT, 1963).

The Second FYDP (1968–1972) was dominated by a tendency to liberalize planning to some extent. The idea of the state entering the housing market as a regulator, not as an investor, was included in this plan. The tendency to decrease the share of housing investments in total investments continued in this plan as well; this time, the rate was reduced to 17.9%. The land and squatter housing policies were generally similar to those of the previous one. The need for low-rent housing built by public institutions and local governments was mentioned in the plan. A new rental law should have been enacted, but there was no proposal or principle on this issue. The difference between this plan and the first one is that it placed greater emphasis on the relations between urbanization and housing policies and gave more importance to the connection between the housing problem and the distribution of the population and settlement policies. During the period covered by this plan, the Land Office was established in 1969. The Land Office would intervene in the market by selling land from the land stock held by the public, but no success was achieved in preventing land speculation, which increased the cost of housing construction, or in land production (DPT, 1967; Çoban, 2012).

The policies proposed in the Third FYDP (1973–1977) were similar to those in the previous plans. In order to follow a more consistent housing policy and to ensure the effectiveness of investments, this plan considered it necessary to combine the housing loan funds provided by various institutions and the resources that individuals can allocate for housing.

Encouraging rental housing for low-income groups, supporting mass housing, and supporting private entrepreneurship along with housing cooperatives were among the priorities set out in the Third FYDP (DPT, 1973).

It is noteworthy that the policies envisaged in the first three plans were also included in the Fourth FYDP (1979–1983). Solving the housing problem with mass housing production instead of individual solutions and directing local governments to build mass housing by providing support to cooperatives were among the priorities of this plan. The use of publicly funded housing loans within integrated policies was one of the principles repeated in the Fourth FYDP. In addition, increasing the funds allocated by social security institutions and public assistance institutions for housing and requiring pension funds to build housing were some of the issues raised by this plan. The allocation of public land to public credit institutions and local government units that construct social housing and placing the land under the control of the public and local governments in cities with the potential for rapid growth was the old policy preferences that the plan reiterated. In addition, it was mentioned that rent control would be implemented. It was emphasized that public housing would be provided for those working in developing regions (DPT, 1979).

During this period, parallel to the FYDPs, municipalities abandoned their reluctance towards building houses. The Innovative Settlements Project in İzmit and the Batıkent project in Ankara were pioneering implementations by municipalities in the housing sector. The Innovative Settlements Project was a project that incorporated many innovations in terms of the way the project was achieved, the organization of demand, construction technology, and the approach of municipalities to the housing problem. The initial stages of the project started in 1973. The project was partially completed in 1990 due to the problems and political changes experienced during the implementation process. The Batıkent project was conceived as an alternative to squatter housing. An area of 10.5 million square metres in the northwest of Ankara was expropriated between 1974 and 1978 with the cooperation of the Ministry of Development and Settlement and Ankara Metropolitan Municipality. This expropriated area was transferred to the housing cooperatives union named Kent-Koop, which was also established under the leadership of the Municipality. The project, which was prepared in 1979, was put into practice in 1980, and the first houses were turned over to the beneficiaries in 1983 (Karasu, 2009; Tekeli, 2012).

As mentioned in the FYDPs, new institutions providing loans for housing were also founded in this period. Established in 1961, the Turkish Armed Forces Assistance and Pension Fund (OYAK) provided individual housing loans and cooperative housing loans to its members. Established in 1971, the Social Insurance Institution for Tradesmen, Craftsmen, and Other Independent Employees (Bağ-Kur) also offered loans to its members. As well as these, the SSK continued to provide loans to its members during this period. The practice of using the funds held by social security institutions in housing finance became widespread in the 1960s and continued until the mid-1980s. Although it was possible to produce housing via long-term and low-interest loans obtained from these institutions, a financing system was not developed for the construction of rental housing for labourers who cannot afford to pay back loans. Thus, the effect this financing system had in solving the housing problem remained negligible (Çoban, 2012).

The end of the 1970s was an unstable period with political turmoil and different governments attempted to formulate different housing policies of their own. In 1979 a decree by the Council of Ministers approved the Policies for Urbanization and Housing Problem through Arrangement of New Urban Settlement Areas prepared by the Ministry of Development and Settlement. These included policies related to the provision of new settlements, land production and distribution, institutional organization, housing construction, home ownership, credit resources, mass transportation, consumer goods, environment, mental health, and production. In addition to these policies, a regulation for providing owner-occupied housing for government officials and a decree concerning the addition of an article to the Law of Banks (Law no. 7129) enabling provision of loans to real and legal persons wanting to build mass and social housing was approved in 1979 (Keleş, 1982).

Mass Housing Law as a Solution to Problems in Housing Market (1980s–2000s)

Towards the end of the 1970s, the inward-looking, foreign-dependent, import-substituting industrialization model that started with the planned period stalled. In order to overcome the bottleneck in the accumulation regime, the foundations of a neoliberal economy were laid with the decisions taken on January 24, 1980, which envisaged integration with capitalist economies. First of all, the import substitution model applied between 1960 and 1980 was abandoned, and integration with the global market became a new target. Minimizing the state's interventions in the economy, limiting the central administration, and transforming the concept of public administration into the concept of public management became the accepted ideas in this period (Karasu, 2009; Çoban, 2012).

In the National Housing Policy decree (dated May 6, 1980), it was stated that due to the rapid population growth and urbanization in Turkey, the cost of housing, sales and rental prices, and the ability to pay of families in need could not be balanced under the current conditions. Considering the definite commitments and principles of the Government Programme in this regard, in order to solve the housing problem as soon as possible due to the insufficient resources and production from the national economy for the housing sector, it was decided to implement a National Housing Policy, which aims at making citizens homeowners and tends towards a balanced and orderly economic and social settlement. In the National Housing Policy

Implementation Principles decree, in order to enable citizens to become homeowners, the following principles were determined:

• Enabling lower- and middle-income groups to become homeowners.
• Expanding the rehabilitation areas in the squatter housing settlements, improving infrastructure services, and seeking practical solutions for land registry opportunities.
• Dealing with housing finance as a whole and making the necessary arrangements.
• Providing large amounts of cheap land in order to reduce the cost of housing, reviewing the standards of all building elements, and giving priority to mass housing and industrial construction in accordance with the National Housing Policy.
• Addressing the housing issue together with the general settlement and urbanization policy.
• Ensuring that citizens working abroad can easily own a home in Turkey.
• The introduction of effective state control over the housing sector.
• Developing housing types suitable for national traditions and customs, taking into account the needs of the Turkish-Islamic family.
• Development of an organization to implement the principles (URL 1).

In parallel with these policies, the 1982 Constitution regulated housing as a right on its own, under the heading of 'right to housing'. According to the Constitution, the state would take measures to meet the housing require-ment. The 1982 Constitution also introduced a new rule that was absent from the 1961 Constitution. In Article 57 of the Constitution, it is speci-fied that the state will take measures to meet the housing needs within the framework of planning that considers the characteristics of the cities and environmental conditions and also supports mass housing enterprises. Thus, the need for shelter is placed in both an urban and environmental context and a planning framework. However, the social class priority of meeting the housing needs of the poor and those on a low income, which was seen in the 1961 Constitution, was abandoned in accordance with the political economy of neoliberal capitalism (Çoban, 2012). State support for the construction of mass housing was an important policy tool in the face of the rapidly growing housing shortage. Therefore, legal arrangements were made in this regard.

A milestone in this period was the Mass Housing Law (Law no. 2487) enacted in 1981. With this law, a fund was established to finance housing production. It was envisaged that 5% of the general budget revenues of this fund would be allocated for housing production. It was aimed to meet the housing need by means of mass housing construction, to determine the prin-ciples and procedures housing developers would be subject to, and to create financial resources to support housing production. The law supported mass production rather than individual housing production and imposed some social limitations on the use of the fund. Accordingly, the person who would benefit from the fund or his/her spouse or children under his/her custody should not already own a house and should be in the lower- or middle-income group. A person would benefit from the financial resources of the fund only once. In addition, the houses to be built within the scope of the law would not be larger than 100 square metres. With the law, for the first time, a privilege was granted to housing cooperatives, and they were given priority in supplying land. However, this law, with its many positive qualities, did not last long in that form. It was amended as a result of political pressure due to the global policies of the post-1980 period and the exclusion of builders who were active in housing production. In 1984 the Mass Housing and Public Partnership Board, the Housing Development Administration (TOKİ) of today, was founded, a new Mass Housing Law (Law no. 2985) was enacted, and the social framework created by the former was significantly damaged. In addition, housing loans granted by social security institutions such as the SSK and Bağ-Kur ended and loans started to be given from the Mass Housing Fund. Despite the negative aspects of this legal change, the Mass Housing Fund is the largest fund in the history of the Turkish Republic, and accordingly, unprecedented amounts of financial resources have been transferred to the housing sector. Financing was provided for housing producers and credit for buyers, and real state support was provided to the housing sector for the first time. During this period, housing cooperatives spread throughout the country due to the loans provided by TOKİ, and the share of cooperatives in housing production increased rapidly. TOKİ both bolstered its financial resources and established a system that works in terms of housing loans, unlike the SSK, Bağ-Kur, and the Real Estate and Credit Bank of Turkey. The main purpose of the law was not to produce housing for low-income families but to spark an economic recovery and to provide the resources needed by the actors in the housing market. A negative side of the process that started with TOKİ was the exclusion of municipalities from the housing supply process. None of the mass housing laws contained any statement about municipalities. Only in the latter one is it stated that the opinion of municipalities would be taken in consideration when deciding where to locate mass housing areas (Karasu, 2009).

FYDPs were also prepared at this time. In the Fifth FYDP (1985–1989), the policies of the government regarding housing were published in an official plan document. Bringing infrastructure to squatter housing areas, improving squatter houses, normalizing ownership, adopting cost recovery for urban development and housing costs, establishing a Mass Housing Fund with non-budgetary resources, and privatization in the housing sector are some of these policies. In this plan, the efficiency of housing sector investments was clearly emphasized and it was stated that in order to obtain the maximum benefit from the economy stimulating and employment creation feature of the housing sector those in need of housing would be encouraged to use savings. In this plan, while it was recommended to reduce the average size of dwellings, it was foreseen that the dwelling units should be considered in reasonable dimensions that conform to the Turkish family structure and lifestyle and reconcile the principles of functionality and economy (DPT, 1984).

In order to achieve housing production as envisaged in the Sixth FYDP (1990–1994), priority in public subsidies was given to people who did not own a house and a decrease in the size of dwellings was required. Supporting the savings funds was among the suggestions. In the plan, it was also suggested that the municipalities would make arrangements to ensure the production of rental and owner-occupied housing for low-income families, assist infrastructure works, and prepare core housing proposals to help people build their own houses in order to prevent squatter houses. This plan also draws attention to the need to establish a link between the investments in the housing sector and the development of cities and settlement purposes. Developing housing construction technologies suitable for the country's conditions, evaluating local construction materials, and giving priority to projects that take into account climatic conditions and reduce waste are among the suggestions of the plan regarding housing. In the plan, it was also suggested that in the construction of housing for public employees, priority should be given to the Priority Regions in Development and to small settlements suffering housing crises (DPT, 1989).

In order to meet the increasing housing needs resulting from rapid urbanization and population growth, the Seventh FYDP (1996–2000) aimed to encourage housing ownership as well as housing production and to develop appropriate financing models that would not impose an additional burden on the public. It was emphasized that institutional arrangements should be made for the use of small savings and the development of the capital market in order to create new resources for housing production. It was aimed to accelerate housing production via projects to be developed in the Priority Regions in Development, especially in the Eastern and Southeastern Anatolian regions. The plan also envisaged the implementation of labour-intensive projects, particularly in the housing sector, in these regions. The policies included carrying out studies for the prevention of environmental destruction and for reducing damage caused by them, supporting and disseminating the developments in housing technology, establishing a Housing Information Bank in order to keep the existing data on housing up-to-date, and conducting housing censuses regularly. In the Seventh FYDP, it was also declared that an amendment would be made to the Squatter Law (Law no. 775) in order to transform squatter houses into legal housing and to transfer the rent acquired to the public, to make self-financing arrangements that allow urban renewal, and to reorganize the provision of land to municipalities (DPT, 1995).

During this period, squatter houses were commercialized and became a means of obtaining rent rather than meeting the need for housing. The main tool in the transformation of squatter housing areas has been the improvement plans and the renewal of these areas and their reintroduction into the urban land market. At the same time, a series of laws were enacted for granting an amnesty for squatter houses (Law no. 2805 and 2981). The legalization and transformation of squatter neighbourhoods were the basic aims of these laws and squatter houses became a constant topic of bargaining between squatters and the government. Urban Improvement Plans were introduced by these laws and squatter houses started to be transformed into blocks of flats through the building and selling system. By the end of the 1980s, urban transformation projects, which were developed with cooperation between the public and private sectors, became an alternative to improvement plans. By several decree laws, mutual or non-reciprocal assistance provided by TOKİ for expropriation, land improvement and arrangement, infrastructure operations, and building construction regarding squatter settlements became possible (Genç, 2014).

Urban Transformation (2000s–Present)

In the 2000s, neoliberal policies became more dominant. With the coming to power of the Justice and Development Party (AKP) in 2002, practices regarding neoliberal policies gained momentum and important legal arrangements were made in the name of state transformation. In this context, important powers have been given to municipalities to produce houses and land and to cooperate with the private sector on this issue in order to obtain a larger share of the rent of housing and land (Karasu, 2009). Furthermore, as public spending on infrastructure and housing projects increased, a boom in the construction sector occurred. Urban transformation became dominant in the national development strategy starting in 2002. As a result, investment in the urban property market became more attractive to industrialists and other sectors of capital. At the same time, the policy response to the global economic crisis of 2008 was to extend efforts towards urban transformation and the urban built environment was financialized. There has been a dramatic increase in public spending on urban transformation projects across the country and on large-scale infrastructure projects that have opened up new fields for the construction industry (Schindler et al., 2020).

After the 2001 economic crisis, the policies supporting the construction industry also led to the acceleration of urban transformation projects and related legal regulations. Squatter houses and squatter neighbourhoods located in the city centres have been where urban transformation has taken place the most. In addition, especially after the 1999 Marmara earthquake in Turkey, the transformation of the urban areas at risk of earthquakes by taking measures to minimize this risk has assumed importance. Urban transformation projects, which were an important item on the agenda of cities in the 2000s, have become an important tool for local governments and regulations changed in the 2000s.

The North Ankara Entrance Urban Transformation Project Law (Law no. 5104), enacted in 2004, can be considered the first example of the new regulations. The purpose of this law is to increase the quality of urban life by improving the physical conditions and environmental image, beautifying, and providing a healthier order of settlements within the framework of the urban transformation project in the areas covering the northern entrance to Ankara and its surroundings. In the Municipality Law (Law no. 5393), enacted in 2005, it is stated that municipalities are responsible for rebuilding and restoring the old parts of the city in accordance with the development of the city and they can implement urban transformation and development projects in order to create housing areas, industrial and commercial areas, technology parks, and social facilities; to take measures to minimize earthquake risk; or to protect the historical and cultural texture of the city. The same statement was made in another law (Law no. 5216), enacted in 2004, with reference to the metropolitan municipalities.

Another law, which came into force in 2005 (Law no. 5366), concerned the renewal, protection, and use of worn out historical and cultural immovable properties and aimed to conduct urban transformation projects involving historical urban textures. The latest law regarding urban transformation was enacted in 2012 (Law no. 6306). Its aim was to determine the procedures and principles regarding improvement, clearance, and renewal in order to create healthy and safe living environments in accordance with the norms and standards of science and art in the areas at risk of disasters. While other laws pave the way for urban transformation projects, it is possible to demolish and reconstruct a single block of flats with this law. With this kind of transformation, the aging housing stock in the planned neighbourhoods located in the central regions of cities was demolished and replaced piecemeal.

A further important development related to the housing policies during this period was the announcement of the Urgent Action Plans by the 58th and 59th governments in the early 2000s. Urgent Action Plans included a countrywide housing programme for new housing production and urban transformation. One of the aims was stated as preventing squatter house construction and transforming the existing ones via planned urbanization. Another aim was to increase owner-occupied housing provision for low-income families through extensive housing construction (Özdemir Sarı, 2019). During this time, TOKİ came to the fore in terms of housing construction. With several laws enacted and with articles added to several laws, TOKİ has become the sole authorized institution for housing and land provision. At that time, although important authority was granted to the municipalities in the field of housing, the central administration did not abandon the idea of a strong central administrative institution, both financially and legally. TOKİ has been given the authority to make plans on any scale and to prepare and implement urban transformation projects. However, no changes were made to the related legislation to increase cooperation with municipalities. Moreover, TOKİ has completely cut its support to housing cooperatives.

No new loans have been given to housing cooperatives since 2003 (Karasu, 2009).

The Mortgage Law (Law no. 5582) enacted in 2007 is another significant development regarding the housing sector in this period is. It aimed to pave the way for home ownership among middle- and middle-low-income groups who cannot buy a house with their own savings and equity capital. Although mortgages are low-interest loans and the repayment period is up to ten years in Turkey, it is essential that the person applying for the loan has a steady income. In this sense, the law, which is considered an important step for the Turkish housing market, focusses on improvements and regulations based on the current situation instead of establishing a new system in the housing market. In today's conditions, mortgages in Turkey cannot provide the necessary conditions for middle-low-income earners, in particular, to benefit from this system (Alkan, 2014).

During this period, along with the new laws and policies, there are four FYDPs. The Eighth FYDP (2001–2005) was prepared so as to stimulate recovery from the devastation caused by the 1999 Marmara and Düzce earthquakes. Increasing housing production and housing ownership was adopted as a principle in the plan. In addition to taking preventive measures against illegal and squatter housing construction, increasing the amount of land with ready superstructures was also among the objectives of the plan. In this plan, unlike the previous FYDPs, the issues of increasing the building and environmental quality in housing production and preserving the historical and natural texture and social and cultural values were included. In the plan, it was also foreseen that the housing finance problem would be solved by institutions that can operate in the capital market and can provide housing loans. The plan also brought up the idea for the establishment of a ministry responsible for urbanization and housing (DPT, 2000).

The Ninth FYDP (2007–2013) did not include any principles related to housing policy. However, a proposal of the former FYDP was implemented in this plan period and the Ministry of Environment and Urbanization was established in 2011 (Decree no. 644).

Policies regarding urban transformation were proposed in the Tenth FYDP (2014–2019). Priority was to be given to transformation projects that would generate high benefits and value in production and common areas, contribute to growth and development, and increase the quality of space and life widely, especially in areas at risk of disasters. It was stated that urban transformation projects would be carried out via an approach that integrated the living spaces of different income groups, reduced workplace/housing distances, was compatible with the historical and cultural heritage of the city, and supported social integration. In urban transformation, ideal area size and integrity would be observed, planning tools would be utilized at the highest level, and procedures and principles that define the qualities of plans and projects, and prioritization, preparation, implementation, monitoring, evaluation, audit, and governance processes would be developed. Models and methods that minimize public expenditures would be used in the financing of urban transformation implementations. According to the plan, in urban transformation projects priority would be given to practices that support innovative and value-added sectors, creative industries, and high-tech and environmentally friendly production. It was also suggested that necessary measures would be taken to meet the basic housing need in the society, especially for low-income groups. Healthy and alternative solutions to the housing problem would be developed. Strengthening the guiding, regulatory, supervisory, and supportive role of the public sector in the housing market was proposed in the plan. Increasing the production of land with ready infrastructure was another policy set out in the Tenth FYDP (TC Kalkınma Bakanlığı, 2013).

The Eleventh FYDP (2019–2023) is still in effect. In this plan, the policy proposals related to housing are stated as follows. Housing needs arising from urbanization, population growth, renewal, and disasters will be met by considering the supply–demand balance. In order to determine the housing need according to the settlements, the housing stock will be determined and during the period covered by the plan, 250,000 social housing units will be built for low-income and disadvantaged groups. The regulatory, supervisory, guiding, and supportive role of the public in the housing market will be strengthened. A coordination mechanism will be established to ensure regular cooperation between units responsible for housing in the public administration. Quality, durability, accessibility, energy efficiency, and disaster resistance standards will be developed and observed at every stage in housing production. The data sources needed in the housing sector will be developed in terms of supply and demand. Housing sector data will be available from a single source in an integrated manner. Research to determine housing demand and content will be increased. In addition, there are also policies regarding urban transformation (TC Cumhurbaşkanlığı Strateji ve Bütçe Başkanlığı, 2019).

Conclusion

As mentioned in detail in this chapter, housing policies in Turkey have been changing in parallel with urbanization and economic policies. In the years when the Republic was founded, the first goal was to provide housing for government officials, but in the second half of the 20th century, the aim was determined as housing for everyone. Although the enactment of the Mass Housing Law in 1984 made a significant contribution to the supply of housing, especially in favour of middle and low-income groups, housing policies were developed in line with the free market mechanism and economic policies. In the 2000s, the urban transformation has become an important housing policy tool for local governments.

The most obvious goal of housing policies has been to increase the number of houses throughout the history of the Republic. Tools such as loan facilities, land facilities, tax reductions, support of building cooperatives, TOKİ applications, and the mortgage system are implemented to achieve this goal. On the other hand, the housing problems of the poor, who need housing and do not have the financial means to buy property, are aggravated. TOKİ has been providing owner-occupied and subsidized housing; however, lifetime subsidies are not provided. Since the establishment of the Republic, the state has not been successful in terms of meeting the right to housing of large segments. Meeting a fundamental right and need has been left to the supply–demand relationship of the market economy. In the housing sector, which is considered one of the main ways out of the capital accumulation crisis, while various components of capital such as lending banks, construction companies, developers, rentiers, and land speculators continue to grow, the housing problem of the low-income segments, both in terms of quantity and quality, remains to the present day. Policies towards owner-occupied housing have not been effective in meeting the housing needs of low-income households in particular. On the other hand, from past to present, housing policies in Turkey have supported home ownership with no effective policy for rental housing.

Note

1 In the building and selling system "… the building plot is given to individual builders or small entrepreneurs for construction of an apartment building. The builder is responsible for obtaining land, supplying the required finance, getting all of the necessary permits, acquiring the building project, and carrying out the construction" (Uzun, 2019: 155).

References

Alkan L (2014) 1980 Sonrası konut politikalarının mekansal yansımaları: Ankara örneği. *İDEALKENT* 5(12):103–131.

Balamir M (1999) Formation of private rental stock in Turkey. *Netherlands Journal of Housing and the Built Environment* 14(4): 385–402.

Çoban AN (2012) Cumhuriyetin ilanından günümüze konut politikası. *Ankara Üniversitesi SBF Dergisi* 67(03): 75–108.

DPT (1963) *Kalkınma Planı (Birinci Beş Yıl) 1963–1967.* Available at: https://www.sbb.gov.tr/wp-content/uploads/2018/11/Birinci-Be%C5%9F-Y%C4%B1ll%C4%B1k-Kalk%C4%B1nma-Plan%C4%B1-1963-1967%E2%80%8B.pdf. Accessed 8 May 2021.

DPT (1967) *İkinci Beş Yıllık Kalkınma Planı 1968–1972.* Available at: https://www.sbb.gov.tr/wp-content/uploads/2018/11/%C4%B0kinci-Be%C5%9F-Y%C4%B1ll%C4%B1k-Kalk%C4%B1nma-Plan%C4%B1-1968-1972%E2%80%8B.pdf. Accessed 8 May 2021.

DPT (1973) *Üçüncü Beş Yıllık Kalkınma Planı 1973–1977.* Available at: https://www.sbb.gov.tr/wp-content/uploads/2018/11/%C3%9C%C3%A7%C3%BCnc%C3%BC-Be%C5%9F-Y%C4%B1ll%C4%B1k-Kalk%C4%B1nma-Plan%C4%B1-1973-1977%E2%80%8B.pdf. Accessed 8 May 2021.

DPT (1979) *Dördüncü Beş Yıllık Kalkınma Planı 1979–1983.*Available at: https://www.sbb.gov.tr/wp-content/uploads/2018/11/D%C3%B6rd%C3%BCnc%C3%BC-Be%C5%9F-Y%C4%B1ll%C4%B1k-Kalk%C4%B1nma-Plan%C4%B1-1979-1983%E2%80%8B.pdf. Accessed 8 May 2021.

DPT (1984) *Beşinci Beş Yıllık Kalkınma Planı 1985–1989*. Available at: https://www.sbb.gov. tr/wp-content/uploads/2018/11/Be%C5%9Finci-Be%C5%9F-Y%C4%B1ll%C4%B1k-Kalk%C4%B1nma-Plan%C4%B1-1985-1989.pdf. Accessed 8 May 2021.

DPT (1989) *Altıncı Beş Yıllık Kalkınma Planı 1990–1994*. Available at: https://www. sbb.gov.tr/wp-content/uploads/2018/11/Alt%C4%B1nc%C4%B1-Be%C5%9F-Y% C4%B1lll%C4%B1k-Kalk%C4%B1nma-Plan%C4%B1-1990-1994%E2% 80%8B.pdf. Accessed 8 May 2021.

DPT (1995) *Yedinci Beş Yıllık Kalkınma Planı 1996–2000*. Available at: https://www. sbb.gov.tr/wp-content/uploads/2018/11/Yedinci-Be%C5%9F-Y%C4%B1ll%C4%B1k-Kalk%C4%B1nma-Plan%C4%B1-1996-2000%E2%80%8B.pdf. Accessed 8 May 2021.

DPT (2000) *Uzun Vadeli Strateji Ve Sekizinci Beş Yıllık Kalkınma Planı 2001–2005*. Available at: https://www.sbb.gov.tr/wp-content/uploads/2018/11/Sekizinci-Be%C5% 9F-Y%C4%B1ll%C4%B1k-Kalk%C4%B1nma-Plan%C4%B1-2001-2005.pdf. Accessed 8 May 2021.

Genç FN (2014) Gecekonduyla mücadeleden kentsel dönüşüme Türkiye'de kentleşme politikaları. *Adnan Menderes Üniversitesi Sosyal Bilimler Enstitüsü Dergisi* 1(1): 15–30.

İçduygu A and Sirkeci İ (1999) Cumhuriyet dönemi Türkiye'sinde göç hareketleri. In: Baydar O (ed.) *75 Yılda Köylerden Şehirlere*. İstanbul: Türkiye Ekonomik ve Toplumsal Tarih Vakfı, 249–268.

Karasu MA (2009) Devletin değişim sürecinde belediyelerin konut politikalarında farklılaşan rolü. *Süleyman Demirel Üniversitesi İktisadi ve İdari Bilimler Fakültesi Dergisi* 14(3):245–264.

Keleş R (1982) Nüfus, kentleşme, konut ve konut kooperatifleri. In: *Kent-Koop Konut 81*. Ankara: Kent-Koop Batıkent Konut Üretim Yapı Kooperatifleri Birliği.

Keleş R (2013) *Kentleşme Politikası*. Ankara: İmge Yayınevi.

Özdemir Sarı ÖB (2019) Redefining the housing challenges in Turkey: An urban planning perspective. In: Özdemirarı Sarı ÖB, Özdemir SS and Uzun N (eds.) *Urban and Regional Planning in Turkey*. Cham: Springer, 167–184.

Özkan A (2009) A critical evaluation of housing co-operatives in turkey within the framework of collective action theories: a case study in Ankara and İstanbul. Unpublished doctoral dissertation, Middle East Technical University.

Schindler S, Gillespie T, Banks N, Bayırbağ MK, Burte H, Kanai J M and Sami N (2020) Deindustrialization in cities of the Global South. *Area Development and Policy* 5(3):283–304.

Sey Y (1998) Cumhuriyet döneminde konut. In: Sey Y (ed.) *75 Yılda Değişen Kent ve Mimarlık*. İstanbul: Türkiye Ekonomik ve Toplumsal Tarih Vakfı, 273–300.

TC Cumhurbaşkanlığı Strateji ve Bütçe Başkanlığı (2019) *100. Yıl Türkiye Planı Onbirinci Kalkınma Planı (2019–2023)*. Available at: https://www.sbb.gov.tr/wp-content/uploads/2019/11/ON_BIRINCI_KALKINMA-PLANI_2019-2023.pdf. Accessed 8 May 2021.

TC Kalkınma Bakanlığı (2013) *Onuncu Kalkınma Planı 2014–2018*. Available at: https:// www.sbb.gov.tr/wp-content/uploads/2018/11/Onuncu-Kalk%C4%B1nma-Plan% C4%B1-2014-2018.pdf. Accessed 8 May 2021.

Tekeli İ (2012) *Türkiye'de Yaşamda ve Yazında Konutun Öyküsü (1923–1980)*. İstanbul: Tarih Vakfı Yurt Yayınları.

URL1 https://www.resmigazete.gov.tr/arsiv/16985.pdf. Accessed 10 June 2021

Uzun N (2019) Transformation in residential areas: Regeneration or redevelopment? In: Özdemir Sarı ÖB, Özdemir SS, Uzun N (eds) *Urban and Regional Planning in Turkey*. Switzerland: Springer, 151–166.

4 Housing and Living Conditions of Turkish Households

What Has Changed in 2000s?

Esma Aksoy Khurami

Introduction

Housing and living conditions include many facets in the representation of a country's household profile. The approaches of studies defining these facets vary. While some focus on the well-being of the household (Bonnefoy, 2007; Busseri and Sadava, 2011; Cracolici et al., 2012; Angel and Bittschi, 2019; Vladisavljević and Mentus, 2019), others underline the importance of housing (material) deprivation (Guio and Maquet, 2006; Navarro et al., 2010; Borg, 2015). International organizations such as the Organisation for Economic Co-operation and Development (OECD), the United Nations (UN), the European Union (EU), and the World Health Organization (WHO) define their measurement method to exhibit housing and living conditions of households and quality of life (Randall et al., 2014). By this means, the OECD has accepted housing conditions as one of the elements of well-being. Housing conditions of Turkish households are reported by the OECD as follows: "The average number of rooms per person has remained relatively stable over the past decade, whereas housing affordability has improved. The percentage of people living in dwellings without basic sanitary facilities has fallen six times more than the OECD average but remains relatively high at 8.2%" (OECD, 2017: 2). However, housing conditions have many more dimensions in addition to those observed by the OECD.

The concern of the EU regarding housing and living conditions differs from that of other institutions. The gap in member states' housing conditions is frequently discussed (Hegedüs et al., 1996) and a common way to measure these conditions is formulated. For this purpose, a dataset called the Survey of Income and Living Conditions (SILC) is developed, which allows cross-country comparison and measurement of material deprivation with the same indicators. This dataset is also developed for countries like Turkey (a developing and EU candidate country). The SILC is conducted within the scope of harmonization processes for membership of the EU. The aim is to generate a dataset enabling comparative analysis of income distribution, poverty, living conditions, and social

DOI: 10.4324/9781003173670-5

exclusion. The SILC is the only and unique dataset to be employed in the investigation of the housing and living conditions of Turkish households with its various variables.

Considering the options in evaluating the housing and living conditions of households, this article approaches Turkish households in terms of their geographical distribution (NUTS Level 1 regions) and modes of tenure. Housing and living conditions are represented by poverty status, material deprivation, housing-related living conditions, adequacy of current income, and the burden of housing and non-housing expenditures. Poverty status is measured both with and without the housing expenditures of households. Material deprivation is examined through affordability to pay rent, mortgage, or utility bills; keep the home adequately warm; face unexpected expenses; eat meat or proteins regularly; go on holiday; and own a television, a washing machine, a car, and a telephone. For housing-related living conditions, housing and environmental problems are queried, i.e., the existence of a leaking roof, damp walls, rotten window frames, heating problems due to poor insulation, dark rooms and lack of daylight, noise from the street and neighbours, air/environmental pollution, and any other issues related to traffic and industry, and crime and violence. In the investigation of the adequacy of current income to meet needs and the burden of housing and non-housing expenditures, the households are subjectively evaluated.

This chapter argues that the housing and living conditions of Turkish households have been affected by the prevalent trends in the 2000s. One of these is the provision of new housing units regardless of the core and peripheral areas of cities. Households started to live in newer and mostly larger housing units. Contractors who built these new housing units have been responsible for all kinds of physical and structural problems and their maintenance in the first five years for newly built housing. Households are expected to spend less money on repairs if they contact the contractors about them. The second trend is the change in the demographic structure of Turkey. Household size has decreased as a result of fewer children. Therefore, housing needs and housing and living expenses have changed over the years. Lastly, housing policies promoting homeownership have moved this mode of tenure to the heart of all sorts of things. Other modes have been neglected and omitted. All these trends have different reflections in the geographies of Turkey.

The study is organized as follows. After the introduction, different approaches in measuring housing and living conditions are reviewed. Housing and material deprivation, the financial burden of housing, poverty, and the inequality dimensions of housing and living conditions are touched on. Then changes in the housing and living conditions of Turkish households are observed based on SILC 2006 and 2018. The last section discusses the current state and sums up the findings.

Different Approaches in the Measurement of Housing and Living Conditions

An overview of housing and living conditions gives clues about the mental and physical conditions of households by highlighting material deprivation, overcrowding, and the burden of housing and non-housing expenditures. Housing and living conditions also appear to influence health (Lowry, 1991; Dunn and Hayes, 2000; Jackson, 2003; Shaw, 2004). Directly or indirectly, indoor air quality and temperature, safety, noise, asbestos, lack of sanitation equipment, and overcrowding are some of the most relevant possible health threats related to the housing itself. To evaluate the experience of the housing and living conditions of households and individuals, two main approaches are apparent in the literature: (i) subjective perception based on the declaration of households and (ii) objective measurement with reference to statistical data. Lee and Marans (1978: 47) interpret objective social indicators as "the description of environments within which people live and work" and subjective social indicators as "the intention to describe the ways people perceive and evaluate conditions existing around them". Alternately, Wolff and Zacharias (2009) explain the objective approach using quantitative parameters measures like income and expenditure, while Diener (1984) describes the subjective approach as a feeling of someone about her/his life as a whole.

Vecernik (2012) comments on the population coverage dimension of the subjective and objective measurement methods. He underlines subjective indicators' applicability to limited sample sizes, while objective ones are applied in censuses and large-scale studies. However, the EU-SILC has become a unique dataset, including both subjective declarations and objective indicators of housing and living conditions such as the burden of housing and non-housing expenditures (subjective declaration) and the ratio of housing expenditures to household income (objective measure). As Sunega and Lux (2016) mentioned, the EU-SILC provides a very limited number of variables for the supplementary use of the objective and subjective measure of housing and living conditions. Although evaluations of housing and living conditions differ between studies, both measurement methods enquire about the ability of households to access goods and services of daily needs, housing conditions, the possession of durables, environmental conditions, and social demographic aspects, which roughly comprise material and housing deprivation, housing well-being, housing burden, and poverty.

Guio and Maquet (2006) determine the housing and living conditions in old and new member states in the EU by employing the 2004 EU-SILC dataset through three different evaluations. Firstly, the financial difficulties experienced by households to meet some aspects of living standards are explored; then, enforced lack of some durables is investigated; lastly, housing-related structural problems are examined. In the cumulative evaluation, they classify households and countries in terms of their level of meeting these living standards. The EU questions material deprivation of households via nine items: "to face unexpected expenses, a one-week annual holiday away from home, to pay for arrears (mortgage or rent, utility bills or hire purchase installments), a meal with meat, chicken or fish every second day, to keep home adequately warm, to have a washing machine, to have a color TV, to have a telephone, to have a personal car" (EUROSTAT, 2011). In that manner, households who lack at least three of the nine items are regarded as living in material deprived conditions.

Mandic and Cirman (2012) and Guio et al. (2009) invoke similar indicators showing provision of a decent quality of life and dignity concerning the physical attributes of dwellings. These components are the perceived lack of space; the presence of rot in windows, doors, and floors; damp and leaks; lack of an indoor flushing toilet; and noise, safety, and lack of open areas in their neighbourhood. They also remark on the difference between the lack of items due to choice and scarce resources. If a household chooses not to have some of these attributes, it is not possible to conclude there is an enforced absence of housing conditions.

Some other studies suggest that it is necessary to extend housing deprivation boundaries to include the economic burden of housing, risks, and security of housing (Palvarini and Pavolini 2010). In this respect, they developed a series of housing deprivation indicators, such as unaffordability, overcrowding, housing inadequacy, quality of the neighbourhood, and tenure insecurity. In this respect, affordability is approached based on the households' perceptions of facing a heavy financial burden of the total housing cost (Palvarini and Pavolini, 2010).

In examining varying housing conditions in continental Europe, Townsend (1979) conducted a pioneering study on poverty and deprivation with a list of 60 indicators covering the material and social needs of the population. Under the housing deprivation heading, structural defects, inadequate amenities, and insufficient space were examined for different sections of society in the United Kingdom. From a narrower perspective, Norris and Shiels (2007) focussed on three aspects: housing quality, accessibility, and affordability. These primary examples of housing condition studies had an impact in different parts of the world. For example, Tao (2015) conducted a study in Beijing and employed housing prices, housing facilities, and housing environment as living condition factors. The size of the house, sizes and number of rooms, space requirements (kitchen, living room, toilets, balcony), housing facilities (electricity, water, internet, security), price (affordability), number of household members (congestion), and other acceptable elements are investigated through a 5-point Likert scale.

With the growth of homeownership in the EU, the focus of studies was partially switched to the relationship between homeownership and housing well-being. In that respect, Filandri and Olagnero (2014) created an index for EU-15 countries' owner and non-owner households using the EU-SILC for 2009. The index consisted of sufficient space per person, location and quality of neighbourhoods and housing, and, last but not least, the excessive burden of maintenance costs on the household budget. If a household gets more than half of the index points, it is regarded as having housing well-being. In comparing the owner and non-owner households in terms of well-being in housing, owner-households are evaluated with higher index scores.

Considering the country-specific context, Niu and Zhao (2018) examined the difference in the housing conditions of migrants in coastal and inland regions of China between 2008 and 2014. Yi and Huang define housing poverty as a lack of optimal housing conditions (private kitchen, private bathroom, natural gas, and heating system) and indicate an association between the lower classes and the more severe housing problems (Yi and Huang, 2014).

As a measure of housing and living conditions, life satisfaction is discussed to assess the current and expected housing and living conditions of households (Pavot et al., 1991). Vladisavljević and Mentus (2019) consider the material and housing needs, income, and unemployment in evaluating housing satisfaction as a reflection of housing and living conditions. Angel and Bittschi (2019) include the significant role of the neighbourhood as a factor in the investigation of the living conditions and satisfaction of households.

Apart from the above-mentioned evaluations regarding housing and living conditions, the burden of housing expenditures influences households in the following ways: (i) increasing housing expenditures lead to a decrease in non-housing spending (Clair et al., 2016; Deidda, 2015; Stone, 2006; Kutty, 2005) and (ii) via negative impacts on the household's health (physical health problems due to inadequate housing conditions and mental stress leading to health problems due to payment difficulties) (Currie and Tekin, 2011). Pittini (2012) refers to and adds an ignored outcome of the housing cost burden problem of households: energy poverty. To decrease this burden, an estimated 52.08 million people in the EU face not keeping their homes warm. In measuring the housing cost burden, the ratio of housing cost to household disposable income (objective) and the self-report by households of the burden of housing expenditures on their budget (subjective) are addressed. In a widespread manner, a household is evaluated as living in affordable housing conditions when the ratio of housing cost to disposable income is below the previously defined score or a household does not declare a burden due to housing expenditures (Maclennan and Williams, 1990). However, the standard of living in housing units is not considered in this measurement.

The housing cost overburden rate is defined by EUROSTAT as "the percentage of the population living in households where the total housing costs (net of housing allowances) represent more than 40% of disposable income (net of housing allowances)" (EUROSTAT, 2014). For homeowners, total housing costs include mortgage interest payments (net of any tax relief), structural insurance, mandatory services and charges (sewage disposal, refuse removal, etc.), regular maintenance and repairs, taxes, and the cost of utilities (water, electricity, gas, and heating), net of housing allowance. For tenants, they include rent payments, structural insurance (if paid by the tenants), services and charges (sewage disposal, refuse removal, etc., if paid by tenants), taxes on the dwelling (if applicable), regular maintenance and repairs, and the cost of utilities (water, electricity, gas, and heating), net of housing allowance. Disposable income, on the other hand, is defined as after-tax disposable household income.

The level of inequality and poverty status are hidden indicators in the definition of the housing and living conditions of households. The disposable income of households represents a way to achieve well-being (Fusco and Dickers, 2008; Cracolici et al., 2012), a frequently used variable to measure these two indicators. To calculate the inequality level in a country, the EU suggests the ratio of the share of income going from the top 20% of the population to the bottom 20% (Atkinson and Marlier, 2010). While there are limited ways to measure inequality, poverty is discussed and measured in many ways, including at-risk-of-poverty rate, residual income-based poverty, and housing-induced poverty.

The at-risk-of-poverty rate corresponds to the share of households whose equalized disposable income is less than 60% of the national median equalized disposable income and it is frequently utilized by economists. On the other hand, residual income-based poverty and housing-induced poverty are commonly employed by housing studying scholars (Freeman et al., 2000; Haffner and Heylen, 2011; Stone, 2006; Tunstall et al., 2013). Housing-induced poverty refers to a situation in which the burden of housing costs makes non-housing goods unaffordable (Kutty, 2005). Along similar lines to housing-induced poverty, the concept of residual income, defined as disposable income minus housing expenditures, allows us to express the non-housing consumption capacity (Hancock, 1993). None of these methods includes judgments about the living and housing standards of households. However, the need to consider housing and living conditions of households as a consequence of the distribution of income (inequality and poverty) is obvious, especially for low-income sections of society.

Observing the Housing and Living Conditions of Turkish Households

In recent decades, several studies have attempted to assess households' quality of life considering housing units and residential environment using nationwide datasets (Pacione, 1990; Bonaiuto and Aiello, 1999; Guio and Maquet, 2006; Norris and Shiels, 2007; Niu and Zhao, 2018). In contrast to the international attempts, studies in Turkey primarily focussed on a city or a neighbourhood unit. The lack of datasets including variables on the housing and living conditions of households and spatial information obliged researchers to perform case studies and obtain data from a limited sample. As one of the first examples of these studies, Türkoğlu (1997) investigated the received quality of the residential environment in planned and squatter urban areas in İstanbul. This was followed by numerous studies focussing on regeneration, urban transformation areas, and mass housing project areas in metropolitan cities such as İstanbul, Ankara, Bursa, and Edirne (Kellekci and Berköz, 2006; Erdoğan et al., 2007; Orhan and Kahraman, 2017; Kahraman and Özdemir, 2017; Sarıoğlu Erdoğdu and Özdemir Sarı, 2018; Gür et al., 2019). While İmamoğlu and İmamoğlu (1992) focussed on elderly Turkish households, Severcan (2019) inquired about children's satisfaction with their residential environment. Although these studies yielded valuable findings, they are limited to their survey areas. Housing and living conditions need to be assessed using various dimensions, including income, quality of life, and material conditions in the country and its regions. To present the existing living and housing conditions of households as the main consequences of policies in recent decades, this study observes Turkey via a countrywide sample through the SILC for 2006 and 2018.

Survey of Income and Living Conditions

In many European countries, the European Community Household Panel (ECHP) was conducted between 1994 and 2001 to determine the living conditions of households at the country level. However, to overcome the challenges in data collection and expand the data collected from households, the ECHP was switched to the EU-SILC. Since 2003, the EU-SILC (cross-sectional and longitudinal) has been a primary data source at the household level to monitor the EU's target topics, such as income, poverty, and social exclusion. The SILC has become regarded as the dataset that should be collected not only by member states but also candidate states within the scope of the EU Compliance Programme. Thus, the Turkish Statistical Institute (TURKSTAT) started to carry out the SILC in 2006. Although it differs in some ways from EU examples, the SILC is the only and primary data source to examine the living conditions of households, especially with geographic reference in Turkey.

It is based on stratified sampling among households registered in the Address Based Population Register System (ADNKS) and the National Address Database (UAVT) in Turkey. These systems also include migrant families from Syria and many other countries. Although it covers a large section of society in its sampling, it lacks some variables such as the legislative status of housing units, rural/urban status of living areas, the amounts of remaining housing debt and monthly housing payment, and the housing unit's price. In the present study, the SILC is employed for 2006 and 2018 to overview the housing and living conditions of households; 10,180 households in 2006 and 24,068 households in 2018 are included from 12 NUTS Level 1 regions.

Housing and Living Conditions of Turkish Households: Empirical Findings

Since the early 2000s, homeownership has been encouraged through many policies targeting every income group in the country. This has led to some

Table 4.1 The changes in the number of rooms, the size of the housing unit, and the size of household

Mean		The number of rooms 2006	2018	Housing unit size (m²) 2006	2018	Household size 2006	2018
	All households	3.41	3.53	97.5	110.0	2.76	2.54
Mode of tenure	Owner-occupier	3.48	3.6	99.6	113.0	2.91	2.68
	Tenant	3.34	3.5	95.6	108.8	2.50	2.35
	Government employee housing	3.19	3.24	89.9	94.1	2.46	2.41
	Others	3.18	3.35	90.6	102.0	2.47	2.32
NUTS level-1 regions	TR1	3.34	3.45	95.8	102.0	2.58	2.54
	TR2	3.31	3.43	89.9	102.1	2.48	2.24
	TR3	3.43	3.40	95.7	100.7	2.64	2.27
	TR4	3.52	3.63	100.4	108.4	2.71	2.47
	TR5	3.72	3.90	101.3	116.6	2.74	2.48
	TR6	3.24	3.39	95.8	113.7	2.63	2.42
	TR7	3.58	3.80	105.9	117.9	2.83	2.54
	TR8	3.47	3.64	100.0	110.5	2.89	2.49
	TR9	3.67	3.78	105.6	111.8	2.74	2.45
	TRA	3.11	3.39	91.6	111.7	3.14	2.92
	TRB	3.60	3.67	103.7	124.6	3.21	2.95
	TRC	3.04	3.27	89.5	114.1	3.01	3.07

Source: Prepared by the author based on the TURKSTAT-SILC (2006, 2018).

changes in the tenure structure of the country. The share of owner-occupiers, which was 65.8% in 2006, declined to 59.4% in 2018. In contrast, the share of tenants (20% in 2006 and 23.9% in 2018) and the "other" category (12.8% in 2006 and 15.3% in 2018) increased. It must be noted that the "other" category here mostly represents households who live in a house owned by their family and pay below-market rent levels or no rent at all. Similar to tenure structure, household and housing features also changed during the observed years. Table 4.1 compares mean values for household size, housing unit size, and the number of rooms in 2006 and 2018 with respect to the mode of tenure and NUTS Level 1 regions. The regions are named as follows: TR1-İstanbul, TR2-West Marmara, TR3-Aegean, TR4-East Marmara, TR5-West Anatolia, TR6-Mediterranean, TR7-Central Anatolia, TR8-West Black Sea, TR9-East Black Sea, TRA-Northeast Anatolia, TRB-Central East Anatolia, and TRC-Southeast Anatolia.

The mean value of the household's size per housing unit was 2.76 people in 2006 and 2.54 people in 2018. This declining trend is supported by a significant increase in single-person households' share from 6% in 2006 to 11.2% in 2018. Household size is observed to decline for all tenure modes and in all regions except the TRC region. In contrast to declining household size, the number of rooms and the dwelling size increased for every tenure mode and in almost all regions. The mean number of rooms increased from

3.41 in 2006 to 3.53 in 2018; the housing unit's size changed from 97.5 m² in 2006 to 110 m² in 2018. These findings highlight the increased consumption of housing space by households in the 2000s, as well as the changing household size and composition.

Based on the existing studies in the field, three main variables are formulated and investigated in the present study in exploring the housing and living conditions of Turkish households: poverty status, material deprivation, and housing-related living conditions. Accordingly, the poverty status of Turkish households is examined first with respect to equivalised total disposable income and then with respect to residual income after housing expenditures are deducted from the equivalised total disposable income. Here, households' disposable income is obtained without imputed rent for owner-occupiers. The poverty line, on the other hand, is set at 60% of the equivalised household income for both 2006 and 2018. Households are categorized into those below and above the poverty line. As Table 4.2 indicates, the share of households facing poverty in terms of both measurement methods decreased from 21.3% in 2006 to 17.4% in 2018, while the percentage of households above the poverty line for both of these methods increased from 71.4 in 2006 to 76.1 in 2018. In other words, the share of households who are at risk of poverty due to income and residual income decreased considerably in the years examined. However, these figures are likely to shift in the opposite direction due to the 2018 economic recession in the country that mostly affected 2019 and the Covid-19 pandemic affecting 2020–2021. These effects are not easy to gauge at the moment.

The separate evaluation of each method reveals that the residual income approach identifies more households as at risk of poverty compared to the total disposable income approach. Moreover, the decrease in the share of households living at risk of poverty between 2006 and 2018 is higher according to the residual income method than according to the total disposable income method. Additionally, the ratio of the total income of the 20% of households with the highest income to that 20% of households receiving the

Table 4.2 Poverty status of households based on equivalised disposable income and residual income

			Equivalised disposable income		
			Yes	*No*	*Total*
Residual income	At risk of poverty?				
	2006	Yes	21.3	6.1	27.4
		No	1.2	71.4	72.6
		Total	22.5	77.5	100
	2018	Yes	17.4	4.9	22.3
		No	1.6	76.1	77.7
		Total	19.0	81.0	100

Source: Prepared by the author based on the TURKSTAT-SILC (2006, 2018).

lowest income (S80/S20) is 8.05 in 2006 and 7.04 in 2018. This means that there is a high level of inequality among the income quintiles, but the gap has decreased in the years analyzed.

In the examination of the material deprivation of households, the approach of the European Commission is adopted. Households' ability to afford the following nine items are questioned: (i) paying the rent, mortgage, or utility bills, (ii) keeping the home adequately warm, (iii) paying unexpected expenses, (iv) eating meat or proteins regularly, (v) going on holiday, (vi) having a television, (vii) having a washing machine, (viii) having a car, (ix) having a telephone. If a household cannot afford at least three of these items, that household is considered to be living in material deprivation. To that extent, the share of households living in material deprivation is shown in Table 4.3. An improvement in material deprivation is observed in all household categories in the years examined. However, the level of improvement and the share of households living in material deprivation varied according to the category of the household. For the tenants and others category, almost 45% of households experienced material deprivation in 2018. For households at risk of poverty (based on the residual income approach), although an improvement was observed in terms of material deprivation, this group still had a higher percentage of material deprivation in 2018. A major improvement in material deprivation is observed for households not at risk of poverty.

Poor housing conditions of households are verified by the existence of at least two out of the following six housing-related problems: (i) leaking roof, damp wall, rotten window frame, (ii) heating problem due to poor

Table 4.3 The share of households experiencing material deprivation and poor housing conditions

| | | The share of households living in | | | |
| | | Material deprivation (%) (lacking at least 3 out of 9 items) | | Poor housing conditions (%) (facing at least 2 out of 6 problems) | |
	Household categories	*2006*	*2018*	*2006*	*2018*
	All households	83.9	37.2	51.6	40.4
Modes of tenure	Owner-occupier	83.2	32.6	48.7	37.7
	Tenant	84.2	45.4	57.1	41.4
	Government employee housing	68.5	13.8	40.6	40.8
	Others	88.4	44.2	58.6	49.1
Poverty status based on residual income	At risk of poverty	95.5	71.5	62.5	53.2
	Not at risk of poverty	79.5	27.3	47.4	36.7

Source: Prepared by the author based on the TURKSTAT-SILC (2006, 2018).

insulation, (iii) having dark rooms and lack of daylight, (iv) noise from the street and neighbours, (v) air and environmental pollution and any other problems related to traffic and industry, (vi) being confronted with crime and violence. Except for the "government employee housing" category, an improvement in housing conditions is observed. The "others" category in tenure modes and households at risk of poverty are identified as the ones experiencing poor housing conditions most, as shown in Table 4.3. Tenants, identified by their high rates of material deprivation compared to owner-occupiers, display similar rates with owner-occupiers in terms of poor housing conditions in 2018. The distribution of households experiencing material deprivation and insufficient housing conditions is also detected at the regional level and displayed in Table 4.4. Although there are declines in the share of households experiencing these problems, still 43.5% of

Table 4.4 The distribution of households experiencing material deprivation and insufficient housing conditions at the regional level

Material deprivation and housing conditions of households			Housing conditions (facing at least 2 out of 6 problems)			
			2006		2018	
			Yes	No	Yes	No
Material deprivation (lacking at least 3 out of 9 items)	TR1	Yes	42.3	35.6	16.1	8.9
		No	9.4	12.7	30.3	44.6
	TR2	Yes	39.0	42.9	13.5	13.5
		No	4.3	13.8	19.2	53.8
	TR3	Yes	45.2	37.7	18.1	19.9
		No	6.4	10.6	14.8	47.3
	TR4	Yes	44.8	34.5	14.2	11.8
		No	5.6	15.1	24.2	49.8
	TR5	Yes	31.6	45.7	10.4	17.6
		No	6.2	16.5	14.5	57.5
	TR6	Yes	54.3	30.0	24.6	16.3
		No	6.9	8.8	20.9	38.2
	TR7	Yes	37.5	48.2	20.3	21.6
		No	4.5	9.9	15.9	42.2
	TR8	Yes	43.2	42.2	17.1	12.5
		No	4.3	10.3	22.9	47.5
	TR9	Yes	44.6	37.8	24.4	15.7
		No	4.7	12.9	22.6	37.4
	TRA	Yes	44.9	44.1	25.9	24.3
		No	3.5	7.5	17.5	32.2
	TRB	Yes	59.8	32.8	25.5	11.1
		No	2.4	5.0	21.9	41.5
	TRC	Yes	72.1	22.7	43.5	22.8
		No	2.1	3.1	12.1	21.6

Source: Prepared by the author based on the TURKSTAT-SILC (2006, 2018).

Table 4.5 The share of households having difficulty and burden

Households having difficulty and burden		Having difficulty living on current income		Having burden of housing expenditures on budget		Having burden of non-housing expenditures on budget	
		2006	2018	2006	2018	2006	2018
	All households	82.6	31.6	87.5	66.9	46.2	55.6
Mode of tenure	Owner-occupier	80.9	27.9	86.7	62.2	42.0	51.6
	Tenant	85.9	37.3	90.3	79.3	58.1	65.3
	Government employee housing	69.9	14.4	72.7	39.0	61.5	58.9
	Others	87.5	38.7	89.4	68.1	47.9	55.5
NUTS level-1 regions	TR1	80.3	22.5	88.9	61.5	44.5	53.2
	TR2	80.5	38.1	84.6	59.5	46.8	52.3
	TR3	80.6	29.1	84.3	64.0	49.1	56.4
	TR4	79.8	37.6	86.2	70.7	55.7	59.7
	TR5	78.7	28.3	85.9	68.8	44.1	61.7
	TR6	84.1	41.6	88.9	65.0	48.5	56.6
	TR7	79.1	27.6	86.4	68.7	36.4	56.9
	TR8	84.5	24.7	88.4	61.1	54.0	58.1
	TR9	77.7	55.8	87.1	77.7	45.6	62.5
	TRA	85.3	27.9	90.7	79.3	40.2	50.8
	TRB	90.5	37.9	88.6	66.9	42.0	51.8
	TRC	94.5	24.1	93.2	72.2	42.8	47.1

Source: Prepared by the author based on the TURKSTAT-SILC (2006, 2018).

households in the TRC region faced both insufficient housing conditions and material deprivation in 2018.

Finally, the burden of housing and non-housing expenditures and the adequacy of current income to meet the needs of households are explored based on the subjective declarations by households (Table 4.5). The heavy burden and slight burden categories in the SILC are combined in the analysis as the presence of a burden. Moreover, without considering the difficulty of meeting needs with current income, if a household indicates any difficulty, it is included in the having difficulty category.

As shown in Table 4.5, for all household categories, the share of households who stated that living on their current income was a problem and they were burdened by housing expenditures declined in 2018. A considerable improvement was seen in the share of households having difficulty living on their current income. Surprisingly, the share of households burdened by housing expenditures decreased between the 2006 and 2018 surveys, whereas the percentage of households burdened by non-housing spending increased. Among the tenure modes, by 2018, tenants were observed to be burdened both by housing and non-housing expenditures in very high proportions.

Discussion and Conclusion

This chapter attempts to examine the housing and living conditions of Turkish households in terms of poverty status, material deprivation, housing-related living conditions, adequacy of current income, and the burden of housing and non-housing expenditures. The results revealed that the housing and living conditions of households improved in 2018 compared to 2006, yet there are regional- and tenure-based disparities in the experiences of households. These findings imply the need to review previously adopted housing policies and to consider policies aiming to promote better housing and living conditions. The current policies are seen to help improve the housing and living conditions of some household groups but not all. The housing policy agenda in Turkey primarily focusses on the households who desire to be homeowners and helps them to achieve this goal. Households who do not want to or who are unable to be homeowners are not a target or beneficiary group of housing policies. However, the study results reveal the need to improve the housing and living conditions of households who are at risk of poverty and "others" and tenant households who are not likely to be owner-occupiers.

The government's recent attempt to construct 100,000 social housing units annually through the Housing Development Administration (TOKİ) to allow low-income households to become owner-occupiers has improved the housing and living conditions of some households. However, households who remain outside the scope of this policy due to the location in which they live do not meet the criteria to benefit from these units. Even though there was an improvement in conditions from 2006 to 2018, 81.6% of the lowest-income tenant households and 72.8% of households in the "others" category belonging to the lowest-income group experienced severe housing and living conditions in 2018. The current housing and living conditions of households could be employed as a criterion in the application and allocation of social housing projects organized by TOKİ rather than using solely the income threshold and residence clause in that city. Moreover, determination of regions that are significantly in need of improvements in terms of the housing and living conditions of households and assessment of the housing needs in these regions can help to enhance conditions.

References

Angel S, Bittschi B (2019) Housing and Health. *Review of Income and Wealth* 65(3): 495–513.

Atkinson AB, Marlier E (eds) (2010) *Income and living conditions in Europe.* Luxembourg: Publications Office of the European Union.

Bonaiuto M, Aiello A (1999) Multi-Dimensional Perception of Residential Environment Quality and Neighborhood Attachment in the Urban Environment. *Journal of Environmental Psychology* 19: 331–352.

Bonnefoy X (2007) Inadequate Housing and Health: An Overview. *International Journal of Environment and Pollution* 30 (3–4): 411–429.

Borg I (2015) Housing Deprivation in Europe: On the Role of Rental Tenure Types. *Housing, Theory and Society* 32(1): 73–93. DOI: 10.1080/14036096.2014.969443.

Busseri MA, Sadava SW (2011) A Review of the Tripartite Structure of Subjective Well-Being: Implications for Conceptualization, Operationalization, Analysis, and Synthesis. *Personality and Social Psychology Review* 15: 290–314.

Clair A, Loopstra R and Reeves A et al. (2016) The Impact of Housing Payment Problems on Health Status During Economic Recession: A Comparative Analysis of Longitudinal EU SILC Data of 27 European States, 2008–2010. *SSM-Population Health* 2: 306–316.

Cracolici MF, Giambona F and Cuffaro M (2012) The Determinants of Subjective Economic Well-Being: An Analysis on Italian-Silc Data. *Applied Research Quality Life* 7: 17–47. DOI: 10.1007/s11482-011-9140z.

Currie J, Tekin E (2011) Is there a link between foreclosure and health? NBER Working Paper No. x17310. Available at: https://papers.ssrn.com/sol3/papers.cfm?abstract_id=1918640. Accessed May 20, 2021

Deidda M (2015) Economic Hardship, Housing Cost Burden and Tenure Status: Evidence from EU-SILC. *Journal of Family and Economic Issues* 36(4): 531–556.

Diener E (1984) Subjective Well-Being. *Psychological Bulletin* 95(3): 542–575. DOI: 10.1037/0033-2909.95.3.542.

Dunn JR, Hayes MV (2000) Social Inequality, Population Health, and Housing: A Study of Two Vancouver Neighborhoods. *Social Science & Medicine* 51(4): 563–587. DOI: 10.1016/s0277-9536(99)00496-7.

Erdoğan N, Akyol A, Ataman B et al. (2007) Comparison of Urban Housing Satisfaction in Modern and Traditional Neighborhoods in Edirne, Turkey. *Social Indicators Research* 81(1): 127–148.

EUROSTAT (2011) 2009 EU-SILC module on material deprivation. Assessment of the implementation. *EUROSTAT Methodologies and Working Papers*. Luxembourg: Publications Office of the European Union. Available at: https://ec.europa.eu/eurostat/documents/3888793/5853037/KS-RA-12-018-EN.PDF. Accessed May 18, 2021

EUROSTAT (2014) Statistics explained. Available at: http://ec.europa.eu/eurostat/statisticsexplained/index.php/Glossary:Housing_cost_overburden_rate. Accessed December 21, 2020.

Filandri M, Olagnero M (2014) Housing Inequality and Social Class in Europe. *Housing Studies* 29(7): 977–993. DOI: 10.1080/02673037.2014.925096.

Freeman A, Kiddle C, Whitehead CME (2000) Defining Affordability. In: Monk S, Whitehead CME (eds) *Restructuring housing systems: From social to affordable housing?* York: Joseph Rowntree Foundation, 100–105.

Fusco A, Dickers P (2008) The Rasch model and multi-dimensional poverty measurement. In: Kakwani N, Silber J (eds) *Quantitative approaches to multi-dimensional poverty measurement*. New York: Palgrave Macmillan, 49–62.

Guio AC, Fusco A and Marlier E (2009) A European Union Approach to Material Deprivation Using EU-SILC and Eurobarometer Data. *Integrated Research Infrastructure in the Socio-Economic Sciences (IRISS) Working Paper Series 19*. Available at: https://ideas.repec.org/p/irs/iriswp/2009-19.html. Accessed May 18, 2021

Guio AC, Maquet IE (2006) Material deprivation and poor housing: What can be learned from the EU-SILC 2004 data? How can EU-SILC be improved in this matter? Draft paper for the conference Comparative EU Statistics on Income and Living conditions: Issues and Challenges, Helsinki. Available at: https://www.stat.fi/eusilc/guio_maquet.pdf. Accessed June 10, 2021

Gür M, Murat D, Şenkal Sezer F (2019) The Effect of Housing and Neighborhood Satisfaction on Perception of Happiness in Bursa, Turkey. *Journal of Housing and the Built Environment* 35: 679–697. DOI: 10.1007/s10901-019-09708-5.

Haffner MEA, Heylen K (2011) User Costs and Housing Expenses. Towards a More Comprehensive Approach of Affordability. *Housing Studies* 26(4): 593–614.

Hancock KE (1993) Can Pay? Won't Pay? Or Economic Principles of Affordability. *Urban Studies* 30(1): 127–145.

Hegedüs J, Tosics I, Mayo SK (1996) Transition of the Housing Sector in the East Central European Countries. *Review of Urban and Regional Development Studies* 8(2): 101–136.

İmamoğlu EO, İmamoğlu V (1992) Housing and Living Environments of the Turkish Elderly. *Journal of Environmental Psychology* 12(1): 35–43. DOI: 10.1016/s0272-4944(05)80295-6.

Jackson LE (2003) The Relationship of Urban Design to Human Health and Condition. *Landscape and Urban Planning* 64(4): 191–200.

Kahraman ZE, Özdemir SS (2017) A Housing Satisfaction Study in an Area of Urban Transformation: The Case of the Türk-İş Apartment Blocks. *MEGARON* 12(4): 619–634. DOI: 10.5505/megaron.2017.04834.

Kellekci ÖL, Berköz L (2006) Konut ve çevresel kalite memnuniyetini yükselten faktörler. *AIZ ITU Dergisi* 5(2): 167–178.

Kutty N (2005) A New Measure of Housing Affordability: Estimates and Analytical Results. *Housing Policy Debate* 16(1): 113–142. DOI: 10.1080/10511482.2005.9521536.

Lee T, Marans W (1978) Objective and Subjective Indicators: Effects of Scale Discordance on Interrelationships. *Social Indicators Research* 8: 47–64.

Lowry S (1991) Housing. *British Medical Journal* 303:838. DOI: 10.1136/bmj.303.6806.838.

Maclennan D, Williams R (eds) (1990) *Affordable housing in Britain and the United States.* York: Joseph Rowntree Foundation.

Mandic S, Cirman A (2012) Housing Conditions and Their Structural Determinants: Comparisons within the Enlarged EU. *Urban Studies* 49(4): 777–793.

Navarro C, Ayala L and Labeaga JM (2010) Housing Deprivation and Health Status: Evidence from Spain. *Empirical Economics* 38: 555–582.

Niu G, Zhao G (2018) Living Condition among China's Rural-Urban Migrants: Recent Dynamics and the Inland Coastal Differential. *Housing Studies* 33(3): 476–493.

Norris M, Shiels P (2007) Housing Inequalities in an Enlarged European Union: Patterns, Drivers, Implications. *Journal of European Social Policy* 17(1): 65–76.

OECD (2017) How's life in Turkey? In: *How's Life?2017: Measuring Well-being.* Paris: OECD Publishing. DOI: 10.1787/how_life-2017-42-en.

Orhan E, Kahraman ZE (2017) Quality of Life in Regeneration Areas: Empirical Findings from the Akpınar Neighbourhood, Ankara, Turkey. *Planlama* 27(3): 314–328. DOI: 10.14744/planlama.2017.75436.

Pacione M (1990) Urban Livability: A Review. *Urban Geography* 11(1): 1–30.

Palvarini P, Pavolini E (2010) Housing Deprivation. In: Ranci C (ed) *Social vulnerability in Europe: The new configuration of social risks.* Basingstoke: Palgrave Macmillan, 26–158.

Pavot W, Diener E, Colvin CR et al. (1991) Further Validation of the Satisfaction with Life Scale: Evidence for the Cross-Method Convergence of Well-Being Measures. *Journal of Personality Assessment* 57: 149–161.

Pittini A (2012) Housing Affordability in the EU-Current Situation and Recent Trends. *CECODHAS Housing Europe's Observatory Research Briefing* 5(1): 1–11. Available at: https://www.portaldahabitacao.pt/opencms/export/sites/ihru/pt/ihru/docs/relacoes_internacionais/CECODHAS_ObservatoryBriefing_Housing_Affordability_2012_revised.pdf. Accessed May 18, 2021

Randall C, Corp A, Self A (2014) *Measuring national well-being: Life in the UK.* Newport: Office for National Statistics. Available at: https://webarchive.national-archives.gov.uk/ukgwa/20160105184137/http://www.ons.gov.uk/ons/rel/wellbeing/measuring-national-well-being/life-in-the-uk--2014/art-mnwb--life-in-the-uk--2014.html. Accessed May 18, 2021

Sarıoğlu Erdoğdu GP, Özdemir Sarı ÖB (2018) Householder Satisfaction in Apartment Block Neighborhoods: Case of Ankara, Turkey. *Journal of Urban Planning and Development* 144(1): 04017022. DOI: 10.1061/(ASCE)UP.1943-5444.0000412.

Severcan YC (2019) Residential Relocation and Children's Satisfaction with Mass Housing. *METU Journal of the Faculty of Architecture* 36(1): 61–84.

Shaw M (2004) Housing and Public Health. *Annual Review of Public Health* 25: 397–418.

Stone ME (2006) What Is Housing Affordability? The Case for the Residual Income Approach. *Housing Policy Debate* 17(1): 151–184.

Sunega P, Lux M (2016) Subjective Perception versus Objective Indicators of Overcrowding and Housing Affordability. *Journal of Housing and the Built Environment* 31(4): 695–717.

Tao LW (2015) Living Conditions—The Key Issue of Housing Development in Beijing Fengtai District. *HBRC Journal* 11(1): 136–142. DOI: 10.1016/j.hbrcj.2014.07.003.

Townsend P (1979) *Poverty in the United Kingdom.* London: Allen Lane and Penguin Books.

Tunstall RM, Bevan J, Bradshaw K et al (2013) *The links between housing and poverty: An evidence review.* York: Joseph Rowntree Foundation.

TURKSTAT (2006) *Survey of Income and Living Conditions-Micro Data Set.* Ankara: Turkish Statistical Institute.

TURKSTAT (2018) *Survey of Income and Living Conditions-Micro Data Set.* Ankara: Turkish Statistical Institute.

Türkoğlu HD (1997) Residents' Satisfaction of Housing Environments: The Case of Istanbul, Turkey. *Landscape and Urban Planning* 39(1): 55–67. DOI: 10.1016/S0169-2046(97)00040-6.

Vecernik J (2012) Subjective Indicators of Well-Being: Approaches, Measurements and Data. *Politicka' ekonomie* 3: 291–308.

Vladisavljević M, Mentus V (2019) The Structure of Subjective Well-Being and Its Relation to Objective Well-Being Indicators: Evidence from EU-SILC for Serbia. *Psychological Reports* 122(1): 36–60.

Wolff EN, Zacharias A (2009) Household wealth and the measurement of economic wellbeing in the United States. Working Paper No. 447. New York: Levy Economics Institute of Bard College.

Yi C, Huang Y (2014) Housing Consumption and Housing Inequality in Chinese Cities during the First Decade of the 21st Century. *Housing Studies* 29(2): 1–22.

5 Limitations of Housing Research Data in Turkey and Proposals for a Better System

G. Pelin Sarıoğlu Erdoğdu

Introduction

The social and behavioural sciences depend upon experimental and observational data to enable increased understanding (Weller and Romney, 1988). Without robust, consistent, and comparable data, housing studies can be limited in many aspects. This issue has been highlighted by many scholars studying housing, especially for non-advanced countries (Igan and Loungani, 2012; Sarıoğlu, 2007; Ge and Hartfield, 2007). The methodologies employed are highly dependent on the available comparable data. Further, in most cases, researchers rely on national data sources that are not specifically prepared for housing purposes. Lack of relevant data can limit the research possibilities (Sarıoğlu Erdoğdu, 2011).

The Western world is known for collecting significantly large amounts of housing data. For instance, EU countries already have housing data accessible to researchers for almost any type of inter-EU country comparisons via the EUROSTAT website.[1] These studies orient the development of housing programmes in those countries. However, in Turkey, the Turkish Statistical Institute (TURKSTAT), the country's primary data provider, has limited data provision for urban studies in general and housing research in particular. This is primarily a reflection of the central administrations' market-based housing perspective in which housing development and maintenance are mostly left to market forces. Without a continuous national housing policy, the need for robust housing research and hence housing data has been neglected to a great extent.

Therefore, in the present study, firstly, several successful examples of data collection and provision for housing research from different countries are given. Then the available data for housing research in Turkey are discussed. In this section, the limits researchers may face are further elaborated. Finally, some proposals are made for a better data system.

International Examples

Western countries are well known for their success in data collection. Housing is no exception in this sense. Among those successful examples,

DOI: 10.4324/9781003173670-6

Australia, EU countries like the Netherlands, and the United States rank top. As an obligation of membership, EU countries collect comparable data that are available through the EUROSTAT website. In those countries, urbanists rely on state-led and freely provided (in most cases) data. On the website, the data on not only member countries but also candidate countries like Turkey are included. Obviously, the data on the Turkish case are dependent on the data provided by TURKSTAT and may not always be comparable to those for other countries.

Australia

The Australian Bureau of Statistics provides housing-related data on topics like population dynamics, housing affordability, finances, homelessness, the rental market, and social housing. In addition to basic variables, specific variables designed for the Australian housing system like 'financial stress', 'wealth distribution', 'financial support', 'waiting-lists for rental dwellings', 'suitability of social housing size', 'tenant satisfaction with services', and 'housing amenities - tenant ratings' are also provided. The data allow comparisons with the other OECD countries as well.[2]

In addition, the website of the Australian Bureau of Statistics has an interactive design that allows researchers to access the latest data and dig into the big data and create customized dashboards. The dashboard includes more than 60 interactive displays with over 8.3 million data points from 25 key national datasets. The dashboard is updated regularly (more than 175 times a year) and new displays are added during these updates.[3]

The housing data are available to researchers through a spatial data catalogue which serves as a central repository for the publication of metadata describing local and state government spatial data. This system enables researchers to search for data, find out what data exist, where and how to access the data, the data's fitness for purpose, who created the data and when and how, how often they are updated, and the geographic extent of the dataset, as well as the rights and restrictions that apply to the dataset.[4]

The Household, Income and Labour Dynamics in Australia (HILDA) Survey is worth mentioning as a relevant example. It is a household-based panel study that collects significant information about economic and personal well-being, labour market dynamics, and family life depending on almost 17,000 families who are followed over the course of their lifetimes. Started in 2001, the HILDA Survey provides insights to enable policy design.[5]

The Netherlands

The Netherlands has a long tradition of large-scale national housing surveys and knowledge-based housing policy (Van Schie, 2006). Accordingly, these surveys were used as a basis for national housing plans and the development of public rented housing stock. Perceiving housing research central to policy development makes the provision of housing data a significant issue in the country.

A Housing Demand Survey (WBO) was carried out within a joint project between the Ministry of Housing, Spatial Planning and the Environment and the Central Bureau of Statistics. The survey (WBO) was conducted every four years between 1981 and 2006 with a minimum of 60,000 households and included almost 700 questions (Sarıoğlu, 2007).[6] The Netherlands' Housing Survey (WoON) replaced the WBO in 2006 and is conducted three-yearly. The survey is a cooperation between the Ministry of the Interior and Kingdom Relations and Statistics Netherlands.[7]

A mixed-mode design is used for collecting the data. Sample units (persons) are asked to participate first via the Internet using computer-assisted web interviewing (CAWI); non-respondents are re-approached using computer-assisted telephone interviewing (CATI) if their telephone number is known and by computer-assisted personal interviewing (CAPI) if it is not.[8]

The WoON Survey provides a huge database for researchers and almost any type of information for housing research. It is not only comparable to that of many EU countries but also allows comparisons within the country. Like the Australian housing datasets, a bulk of questions unique to the Dutch housing system (like the mortgage and public rented housing) are included. Moreover, information on previous dwellings and expectations for future homes are available (Sarıoğlu, 2007). Such information is relevant in the sense that it enables better design and planning of housing policies.

The minority population is significant for the survey, and questionnaires are available in other languages like Turkish (Sarıoğlu, 2007). This is valuable for researchers studying the Dutch housing system in other languages. Furthermore, WoON enables urban/rural differentiation on a scale of 5 and uses population density rather than the total population. This is a better classification method for urban analysis than dual urban/rural differentiation (as in the Turkish case until 2014).

The USA

Another successful example is the USA's American Housing Survey, carried out by the Census Bureau and the Housing and Urban Development. The survey has been conducted with a predetermined and fixed set of 50,000 households every two years since 1985, providing retrospective housing and household information. Moreover, for new construction and for some metropolitan areas, the survey is carried out with additional samples.[9] The basic advantage of these data is their fixed set of households, as retrospective household and housing information can be a basis for further housing studies.

In the United States, additionally, the Panel Study of Income Dynamics (PSID), which provides valuable information to housing researchers, has been carried out since 1968. Although not directly prepared for housing research, its demographic and sociological content makes it valuable for housing studies. Its sample size increased from 4800 to 7000 households in 2001.[10]

A positive aspect of the data collection is that an advisory board of researchers, policymakers, and scientists has guided the data collection process since 1982. This is a valuable input strengthening the bonds between research and policymaking.

Until 1972, face-to-face interviews were used to collect data, but with technological improvements, after 1972, the telephone was used as a primary method of data gathering. Since 1993 computer-aided surveys have been implemented.[11] These methods accelerate not only the data collection process but also the digitization of the data.

The Turkish Case

In contrast to the successful international examples, there is no continuous housing data provision in Turkey specifically designed for housing research. Balamir (1985) was one of the first authors to point out the absence of housing data in Turkey and the problems and difficulties in using housing-related datasets instead.

Among a number of sources of housing-related data, many are subject to change and are not continuous (Sarıoğlu, 2007). Scholars have to follow up different data providers and datasets, which in most cases provide indirect housing information. A systematic data provision that is easy to access and combine with other datasets when necessary is lacking (Ministry of Development, 2018).

Another major problem is the geographical coverage of data. It does not always allow different spatial levels of analysis in housing research; in particular, city-based studies are almost impossible with many of the datasets. Further, conventional survey methods are usually employed, with underuse of the telephone and Internet in most cases (Sarıoğlu 2007).

In Turkey, the absence of specific housing data primarily stems from the general understanding and perspective of the central administrations over the years. Housing has never been seen as a 'basic need' in Turkey, where governments have to play a major role in the allocation and maintenance of housing units (Sarıoğlu Erdoğdu, 2013). Thus, no major emphasis has been given to the provision of housing data in the country.

The lack of adequate statistical databases is immediately noticeable, especially when making cross-country comparisons. Social and cultural aspects have been neglected not only in housing provision but also in housing data collection. The absence of these aspects in data provision is so dominant that its implicit impacts on studies are inevitable (Sarıoğlu, 2007).

In Turkey, some public institutions provide various data related to housing research. Current data provisions are scattered so that finding the right and appropriate data for housing studies can be a tough and time-consuming job. The following grouping of housing data providers can be helpful in this sense:

1 TURKSTAT
2 Ministries (Ministry of Environment and Urban Planning, Ministry of the Interior, Ministry of Finance, and Ministry of Agriculture and Forestry)
3 Municipal records
4 Other sources (Central Bank (CBRT), Union of Chambers and Commodity Exchanges of Turkey, and Natural Disaster Insurance Institution).

TURKSTAT

TURKSTAT is the primary data provider in Turkey. TURKSTAT has been collecting data at many levels and in many ways, but only some of them can be used in housing research. In the special expertise commission report on Housing Policies prepared by the former Ministry of Development (2018), it was stated that the main challenge for housing research in Turkey is that special datasets are not designed, and existing ones (for example, TURKSTAT Population and Housing Research (NKA)) are not continuous. The NKA, carried out in 2011, provides provincial-level information on the workforce, employment, and unemployment, the reason for migration, disability, and features of buildings and housing units, which cannot be obtained from the National Address Database (UAVT) (Ministry of Development, 2018).

Housing datasets used in research in Turkey are mostly micro datasets designed by TURKSTAT for different purposes (Ministry of Development 2018). Scholars in Turkey use two of these micro sets: The Household Budget Survey (HBS) and Income and Living Conditions Survey (SILC). A third dataset available from TURKSTAT is the Housing and Construction Statistics.

Household Budget Survey

The HBS, conducted by TURKSTAT, is one of the primary sources urbanists rely on. The survey aims to obtain information on the consumption patterns of households, and housing expenditures are included among the consumption items in the survey. The earliest survey covering the whole of Turkey was carried out in 1987 and then in 1994, but are not comparable with those conducted after 2002 as the spatial coverage is different (Sarıoğlu, 2007). Since 2002 annual regular surveys were launched including a total of monthly 800 and annually 9600 sample households. The sample size was increased to 1296 households monthly and 15,552 households yearly in 2009 (TURKSTAT, 2015).

These data allow comparable studies with EU countries as the same EU methodology and recommendations for harmonization are used to determine questionnaires and survey methodology by considering country conditions. Face-to-face interviews with members of households and by diaries in which households record their daily expenditures during a survey month are used in obtaining data.

The HBS enables researchers to study the building date, number of rooms, housing area, the rental value of the house, imputed rent, house value declared by the landlord, type of ownership, household size, household type, household income, length of residence, and household housing expenditure and sub-breakdowns (maintenance, water, electricity bills, etc.) (Ministry of Development, 2018).

Although the range of the variables is promising, as the geographical reference is the whole of Turkey, the data do not allow analysis in sub-spatial scales like regional, provincial, and urban areas (Ministry of Development, 2018). Moreover, one can argue that the unavailability of retrospective housing information and future residential expectations of households, like in the WBO in the Netherlands, for instance, is a disadvantage in comparative studies. Further, the WBO is conducted in other languages like Turkish, whereas the HBS is not carried out in other languages, disregarding the residential information of the migrant population (Sarıoğlu, 2007).

Income and Living Conditions Survey

Another data source provided by TURKSTAT is the Income and Living Conditions Survey (SILC), which covers the period since 2006. Before 2006, independent surveys on income distribution were available for 1987 and 1994 from household budget surveys (TURKSTAT, 2020a). The absence of panel data was made up for in this sense.

Like the HBS, this survey is also prepared within the scope of studies' compliance with the EU. The survey aims to supply comparable data on income distribution, living conditions, social exclusion, and relative poverty. Estimates are produced on Turkey, urban and rural areas, and NUTS Level 1 (covering 12 territorial units) from the annual cross-sectional data and two, three, and four years' panel data for the whole country (TURKSTAT, 2020a). NUTS Level 2 (covering 26 territorial units) information has been produced since 2014[12] (TURKSTAT, 2020a).

The SILC has many variables that can be used in housing research including the type of ownership of the house, the rental value of the house, the imputed rent, NUTS Level 1 regional codes, household income, subjective evaluations of the households regarding housing problems, and their housing expenditure burden (Ministry of Development, 2018). Again, city-based housing research is not possible with SILC data, except for the three major provinces, namely Ankara, Istanbul, and İzmir. Yet, these exceptions still do not represent the urban population. The sample size of the survey has been increased gradually since 2014 to produce estimates on NUTS Level 2.[13]

Housing and Construction Statistics

Thirdly, TURKSTAT provides data on building and occupancy permits which are used by housing researchers to indicate the number of dwellings in terms of house type, floor area, and value. Building and occupancy permits are two of the oldest data sets available and information up to 2007 can be accessed through the TURKSTAT website. Between 2007 and 2014, data could be obtained from the Ministry of the Interior (TURKSTAT, 2020c). These building documents provide information on address, building permit information, title deed information, the identity of the owner of the building, general characteristics of the building (surface area, number of independent sections, number of floors, and dwelling units and their types), technical features of the building (heating system, fuel type, water/waste connection status, hot water supply, the parking lot, the coal store, the elevator, the shelter, the pool, the structural system of the building, and material information regarding the walls and floors) (Ministry of Development, 2018). Although the variable list is long, not all are available to researchers due to privacy concerns. Variables on technical features and general characteristics of the building are easily accessible on the TURKSTAT website.

Likewise, house sale statistics provided by TURKSTAT is another data source for housing researchers announced monthly. Since 2013, these data geographically cover the whole of Turkey and include information not only on house sales with mortgages but also on first sales, second hand sales, and house sales to foreigners (TURKSTAT, 2020b). The data are not provided in raw format and it is not possible to carry out case-based studies with this dataset. As the data source is the database of the General Directorate of Land Registry and Cadastre and Turkey's National Geographic Information System, the information available might be unreliable as the prices are, generally speaking, below market values. This issue is further discussed in the following sections.

A recent development was the introduction of the Official Statistics Program by TURKSTAT, which aimed to implement a systematic approach. In a report prepared for the 2017–2021 term, it was stated that datasets obtained through surveys and similar methods had been replaced by administrative registers in terms of producing fast, economical, and consistent statistics in many developed countries such as France, Denmark, Finland, Norway, Sweden, the United States, Canada, Australia, and New Zealand (TURKSTAT, 2017).

According to the report, regarding both the national priorities (Government Programme, Priority Transformation Programme, 10th Development Plan, and related strategy plans) and the international obligations (closing criteria of the EU Statistics Chapter, data transfer to international organizations), the use of administrative records for statistics becomes a priority (TURKSTAT, 2017).

In this context, a number of registration systems have been developed, such as the UAVT, Address Registration System (AKS), Business Registration System, Agricultural Enterprise Registration System (TİKAS), and Migration Registration System (Göç-Net) (TURKSTAT, 2017). The UAVT and AKS registers promise to form a housing database, especially if linked to spatial data (TURKSTAT, 2017).

Ministries

Several ministries in Turkey collect data regarding their interests, intervention areas, and priorities. Among those datasets, housing researchers can use ones like the Ministry of Finance Revenue Administration's Rental Income Statements of households. Some of those registries are mentioned in the following sections.

Ministry of Environment and Urbanization

Working under the Ministry of Environment and Urbanization, the General Directorate of Land Registry and Cadastre (TKGM) officially keeps data related to all buildings and land. The TKGM keeps the data not only in hard copies (sheets, registers, official bills, etc.) but also in digital format: (1) Land Registry and Cadastre Information System (TAKBİS), where land registry information is kept and (2) Spatial Real Estate System (MEGSİS), where cadastral information is kept. The data saved in these systems are shared online with many public institutions and organizations.[14] Further, the plot information for all real estate throughout Turkey is publicly available via an online system.

The TAKBİS data have a number of deficiencies, however, like the unstandardized data collection (e.g., categorical classification of dwelling, housing, and house) and digitization, not being able to transfer some changes (as in the current status of the buildings) to records quickly, and unreliable information (as in the purchase price of the dwelling, which is generally below the market value) (Ministry of Development, 2018). A house price index could not be developed in this sense, and for now, the current index does not represent the whole of Turkey as it only includes data from expertise firms.

Ministry of the Interior: National Address Database (UAVT) and Spatial Address Registration System (MAKS)

The UAVT, developed by the Ministry of the Interior, also provides information to housing researchers, like the number of dwelling units and workplaces available at the neighbourhood level.

Depending on this database, a recent project as a part of the Information Society Strategy 2006–2010 Action Plan[15] was initiated: the Spatial Address Registration System Project (MAKS). It is a system established in order to combine the text information contained in the AKS with geographic coordinates and to integrate the created infrastructure into other systems.

MAKS aims to increase efficiency in public services, allow more accurate analyses when planning services and investments, and enable shorter response times in the event of disasters, fires and other emergencies, and public order incidents. At the beginning of 2020, information on almost 45% of the population and 35% of the country's area was transferred to the system.[16]

Through MAKS, researchers can access information like building type (public, residential, workplace), the number of floors (underground and above), independent section use (residence, hotel, nursing home, student dormitory, workplace, school, etc.), independent section ownership type (private/public), independent section area (from building documents), and the coordinates/locations of addresses (Ministry of Development, 2018).

When completed, this project may be very promising for urbanists and, specifically, housing researchers, although it is not certain how specific the data, scale, periods, and estimation levels will be.

Ministry of Agriculture and Forestry

The CORINE (Coordination of Information on the Environment) Land Cover Project provides the same basic data for the purposes of determining environmental changes in the land, rational management of natural resources, and determining environmental policies in all member countries of the European Environment Agency in line with the criteria and classifications set by that agency. It was started in 1985 to collect data and create a standard database, and so far, datasets for 1990, 2000, 2006, and 2012 have been produced (TURKSTAT, 2017).

Using remote sensing and geographical information systems, satellite images, and aerial photographs to support traditional data collection methods, statistics related to artificial and natural areas (residential areas, roads, agricultural areas, forests, water bodies, soil erosion, land structure, natural disasters, pollution, etc.) can be produced depending on the location (TURKSTAT, 2017).

The geographical coverage of the CORINE project is the whole of Turkey. By using remote sensing and geographic information system technologies, computer-aided visual interpretations are made of medium- and high-resolution satellite images like Landsat-4/5/7, TM, SPOT-4, and IRS LISS III. These images are provided by the European Environment Agency. The information is available at 1/100,000 scale developed by using 1/25,000 scale topographic sheets, soil, forest stand, and irrigated area maps (TURKSTAT, 2017).

CORINE helps in the analysis of changes in land use, which can be crucial for housing research. However, the scale of the data prevents micro-
level analysis. Further, household characteristics cannot be linked to these
data, limiting research possibilities.

Municipal Records

According to the Real Estate Tax Law, all buildings and land within the
country's boundaries are subject to building tax and land tax. This information is comprehensive in the sense that it covers all the land and buildings.
The basic problem with this information is that the values determined as tax
bases are far below the real estates' market values and this method might
be very different from scientific and international standards (Ministry of
Development, 2018).

Municipalities keep information on many topics and Real Estate Tax
Records are one of them. These records are the statements made by the
taxpayers through the real estate tax notification form and the information in the title documents submitted to the municipality during this declaration. In this context, municipal property registers contain information
like taxpayer identification and contact information, neighbourhood,
street, outside/inside door no. and title deed, plot, island/parcel, volume/
page no. information, building plot size, type of construction (steel, rein-
forced concrete, masonry, wood, stone wall, slum, etc.), construction class
(luxury, first, second, etc.), current use (residence, workplace, warehouse,
factory, hospital, etc.), and construction completion date (Ministry of
Development, 2018).

Other Data Sources

Some other public bodies like the Union of Chambers and Commodity
Exchanges of Turkey (TOBB) and Central Bank make available housing-related data. TOBB provides disclosing information about companies, associ-
ations, and natural entities as Established and Closed Companies' Statistics,
which TURKSTAT had supplied until 2010.[17] The data include information
like the type of companies and foreign partnerships, which could be valuable to housing researchers.

The construction of a house price index (HPI) for Turkey had long been
one of the considerations of the CBRT (Kaya et al., 2013). The Central
Bank's Housing Price Index has been produced monthly since 2010, aiming
to reveal price differences in house sales. It is based on the house credit
(mortgage) information of the real estate value assessment reports. The data
are available for 81 provinces, which means urban/rural differentiation is
not possible.[18] As a major drawback, housing sales through the Housing
Development Administration of Turkey are not covered in the index and the
data constitute less than 50% of the houses sold. Özdemir Sarı (2019: 173)

found that the data employed in the calculation of the HPI could hardly cover 35–40% of the dwellings in the housing market that are for sale. Hence, the index does not adequately represent the housing market (Ministry of Development, 2018).

Another useful source for housing researchers could be the Natural Disaster Insurance Institution (DASK), from which information on Compulsory Earthquake Insurance Records can be accessed. Ownership ID (Turkish Republic ID number of the individual), building style (steel/reinforced concrete, masonry, and other), building construction year (1975 and before, 1976–1996, 1997–1999, 2000–2006, 2007, and later), and residential area information is compiled (Ministry of Development, 2018).

Thus, there are numerous sources of housing data in Turkey and yet housing research still suffers from multiple problems related to data. The next section outlines the major drawbacks regarding the data available.

Problems and Proposals

Problems

The data available seem to be extensive when listed, but scholars studying housing in Turkey at almost all spatial levels suffer from unsystematic and unstandardized data collection and provision as a major issue. Sometimes it is the legal and technical difficulties in sharing and merging data between different bodies, and sometimes it is the unstandardized data collection and differences in data definitions that prevent systematization (Ministry of Development, 2018).

Available data are subject to change, and, further, they are not available from a single source. In addition, regarding the protection of personal data, some of the data cannot be shared with researchers. It is time consuming for scholars to search for the right data and access them even before the real analysis starts.

Difficulties in housing research, hence, can be grouped as the total absence, discontinuity, inadequacy, and incomparability of housing data. A final problem relates to the definition of urban areas by TURKSTAT.

Total absence: The data needed does not exist at all. Retrospective housing information, for instance, is one example. Information on previous dwellings, reasons to move in and out, and moving destinations are not available. Hence, research on the mobility of households and the relations between life-cycle events and housing choices can be almost impossible to carry out. Depending on the duration of stay at particular dwellings, mobile households could be determined indirectly.

In such cases, researchers have to pursue their research interests through scientific projects so that specific surveys can be carried out. Yet, individual surveys may be limited in terms of spatial coverage and/or representability due to the time or budget constraints of the projects. Further, they are used only by those involved in the project and are not used in future studies to a great extent.

Discontinuity: Available data may change in scope and design. For instance, during the EU transition process, the spatial level of the datasets in Turkey was changed from urban/rural differentiation to the NUTS system. Such changes might open up ways for comparability with EU countries and yet also compel researchers to spend additional time to learn the details of the new datasets, which means an additional workload for them. Specifically, for urban planners, the NUTS system does not allow urban/rural comparisons and city-based studies, which are crucial to the discipline. Information on the Housing Price Index of the Central Bank of Turkey is an example of this. The data are appropriate for provincial studies on the three biggest cities and regional research on the other cities in the country.

Similarly, as mentioned above, the NKA (2011) is one of the only housing data provisions, allowing only cross-sectional analysis. All the efforts to prepare and publish the NKA became worthless in just a few years, as most researchers and policymakers clearly preferred to work with the latest data unless they had a specific interest in 2011.

Inadequacy: Even when there is some information that might be useful to housing researchers, it comes from indirect sources that are not specifically designed for housing research. For instance, the population census used to be the major source of determining tenure type before 2000. Indirect sources are inevitably inadequate for enlarging the research and using the most appropriate method.

The HBS and SILC have some potential as the two major data provisions, and yet they are inadequate for many housing research areas as firstly, there is no way to conduct city-based housing analysis, and secondly, they are not connected in terms of variables and household samples to enhance the research. The availability of panel data does not enable longitudinal studies as they are not applied to the same households with the same variables. Further, the HBS and SILC have been conducted since 2002 and 2006, respectively.[19] They are relatively new datasets, especially when considering other examples (the WBO, for instance, has been conducted since 1981).

Not only the HBS and SILC but also most of the data presented are inadequate in terms of geographical reference. In most cases, datasets enable regional level analysis, inadequate for urban studies. The geographic reference of the datasets should be extended to cover urban and even neighbourhood levels for housing research.

Incomparability: Comparative studies of great importance in housing research are as old as housing research itself (Ball et al., 1988). Further, it is relevant to see how other countries manage their housing systems, and this is why politicians and researchers are curious in international housing research (Ball et al., 1988). As a result of globalization in general and the process of enlargement of the EU in particular, international comparative research has gained impetus among many European scholars (Sarıoğlu Erdoğdu, 2011).

However, the datasets TURKSTAT provided do not extend over time and space, preventing retrospective and spatial comparisons in most cases. This is true not only for inter-city Turkish comparisons but also for country-based comparisons.

The HBS, for instance, is available back to 2002 and could be used as a major source for a longitudinal study. Yet, the urban/rural definition and methodologies have changed such that information for 2004–2013 is not comparable with that for after 2014.[20] The law numbered 6330 on Metropolitan Municipalities, which has been in force since the March 2014 local elections, plays a significant role in this as the central government changed administrative borders afterwards. Villages gained the status of neighbourhoods in some provinces in this process. Making urban/rural distinctions depending on official records became meaningless in some ways.

Urban/Rural Definition: Housing studies can be carried out at different scales. Urban levels and, specifically, city-based studies are of great importance to urban planners. Current data provisions do not allow city-based studies or urban/rural distinction, which is becoming a major issue.

In a report by TURKSTAT it was declared that since 2014 the rural and urban distinction is omitted from the statistics supplied by TURKSTAT. All the data have been produced to cover the whole of Turkey. This decision was made based on an argument that the urban/rural definitions used by TURKSTAT lead to the production of biased estimates for the country as a result of neglecting the effect of the rural structure or urban structure on the general representation (TURKSTAT, 2017).

During the handling of a significant problem of statistics, another problem was created by this urban/rural definition. As such, many scholars are forced to carry out their own surveys in order to conduct housing research at the targeted scales.

Proposals

Regarding the problems and difficulties, some necessary proposals can be made to allow better housing data and housing research in Turkey. They can be listed under three headings: *The system (coordination and systematization), data collection (methods, data reliability, and quality), and access to data (sharing and updates).*

The System: Coordination and Systematization

Firstly, the systematization of the data provision is a must in the Turkish case. Working with TURKSTAT from the very beginning, an effective coordination between institutions should be ensured (Ministry of Development, 2018). The introduction of the Official Statistics Programme (RIP) may be regarded as a beginning for such a central database and systematization. And yet, after three years, the expected results have still not been achieved. During the systematization process, a key issue could be to have a hierarchical structure which delegates some of the responsibilities to certain bodies while keeping the main control at TURKSTAT.

From the perspective of researchers, priority should be given to easy access from one particular source. The system should be readable by all users and developed such that it enables dynamic control over the variables.

Data Collection: Methods, Data Reliability, and Quality

The design of a housing survey that is continuously conducted is the first step towards overcoming the data-related housing research problems in Turkey. In doing so, scientific methods and international standards must be employed for collecting data (Ministry of Development, 2018). This is not to say that current datasets are collected and provided in unscientific ways. Rather, the intention is to point out the need for quick adaptations from theory and updates from international standardization developments. Only afterwards can internationally comparable and robust housing data become available.

Regarding data collection, another proposal denotes the absence of inadequacy of spatial linkages in the available datasets. Any housing related dataset should be linked to MAKS, which was suggested in the 11th Development Plan Housing Special Commission report as well (Ministry of Development, 2018). With no geographic reference, most of the data collected by TURKSTAT become irrelevant for housing researchers.

Collecting data is a time-consuming and expensive process, and once the survey is carried out, it is almost impossible to make further changes (Sarıoğlu, 2007). Hence, the design of data collection should be carefully considered. An advisory board for TURKSTAT studying alternative models of data collection would be helpful for guiding and orienting the process in this context.

In order to increase data quality and reliability, an independent supreme board with advisors from public and private bodies and researchers from different disciplines can provide insights as in the USA PSID data. Such a proposal was also made in the 11th Development Plan Housing Special Commission prepared by the Presidency of Strategy and Budget (Ministry of Development, 2018).

Further, data collection methods can be increased and supported by new technologies like the use of telephone and online surveys. TURKSTAT has started to employ some of these methods only recently and they need to be expanded.

Access to Data: Sharing and Updates

After targeted housing becomes available, further arrangements are needed. The data should be easy to access and use and updated simultaneously when new data become available. Sometimes the data are not provided free. Yet, when research is conducted for graduate level studies, for instance, the fees are waived. This could accelerate the process for research.

Electronic and online provision of data via a central source and/or integration of databases among different institutions is also argued to be significantly affecting housing research (Ministry of Development, 2018). In this sense, legal arrangements could be necessary in order to ensure inter-institution data sharing. In addition, a number of further datasets can be made available for research if alternative ways of data presentation are developed that ensure the protection of personal data.

Concluding Remarks

The aim of this study was to analyze the available datasets which are widely employed in housing research in Turkey. As the categorizations used in the Problems and Proposals section reveal, they are mostly data provisions produced by public bodies in accordance with their priorities. As none of those institutions consider housing a primary issue, all the available data could be argued to be indirect for housing studies. Hence, housing research in Turkey suffers from the absence of a specific housing data provision. The general understanding which underrated the relation between housing research and successful policymaking over the years is responsible for the current situation (Sarıoğlu, 2007).

In fact, an understanding of the housing sector would be very valuable to policymakers as many relations in the urban space are affected significantly by changes in this sector. Only within this perspective could increasing the amount of research in the housing sector and creating consistent and comprehensive data sources be possible (Balamir, 1985).

It is difficult to use indirect raw material in research, and the absence of housing data in Turkey is a bigger problem than it seems. Without changing the dominant understanding that housing is just a part of the construction industry, real changes cannot be achieved for housing scholars.

Obviously, a few suggestions might be made for a better system. Proposals developed could pave the way in this sense. In addition, some international obligations such as those involved in the EU integration process and standardization policies are enforcing data collection and provision so that comparable housing studies may be carried out at least at the national level.

The final remark relates to the need for establishing a better understanding of housing research not only in the eyes of policymakers and central agencies but also for some scholars in Turkey. The significance of linking housing research to policy development will eventually reveal that many other sectors and individuals are financially, spatially, and socially affected by even small changes occurring in the housing sector. Only if this understanding gains impetus will successful housing data provision become possible for Turkish housing researchers.

Notes

1 https://ec.europa.eu/eurostat/home? Accessed September 28, 2020.
2 https://www.housingdata.gov.au/ Accessed September 28, 2020.
3 https://www.housingdata.gov.au/ Accessed September 28, 2020.
4 See NSW Spatial Data Catalogue on https://data.nsw.gov.au/data/dataset/ sydney-region-dwellings Accessed September 15, 2020.
5 https://melbourneinstitute.unimelb.edu.au/hilda Accessed March 14, 2021.
6 https://www.cbs.nl/en-gb/our-services/methods/surveys/korte-onderzoeks-beschrijvingen/netherlands-housing-survey--woon-- Accessed April 10, 2020.
7 https://www.cbs.nl/en-gb/our-services/methods/surveys/korte-onderzoeksbes-chrijvingen/netherlands-housing-survey--woon-- Accessed September 15, 2020.
8 https://www.cbs.nl/en-gb/our-services/methods/surveys/korte-onderzoeks-beschrijvingen/netherlands-housing-survey--woon-- Accessed April 10, 2020.
9 http://www.huduser.org/datasets/ahs.html AHS Questionnaire is accessible via https://www.census.gov/programs-surveys/ahs/ Accessed March 27, 2020.
10 https://psidonline.isr.umich.edu/Guide/ug/psidguide.pdf Accessed March 27, 2020.
11 https://psidonline.isr.umich.edu/data/Documentation/UserGuide2017.pdf Accessed March 13, 2021.
12 https://data.tuik.gov.tr/Search/Search?text=silc&dil=2 Accessed March 14, 2021.
13 https://data.tuik.gov.tr/Search/Search?text=SILC Accessed May 12, 2021.
14 https://www.tkgm.gov.tr/en/page/land-registry-and-cadastre-information-system-takbis Accessed September 23, 2020.
15 https://maks.nvi.gov.tr/ Accessed March 29, 2020.
16 http://www.ankara.gov.tr/vizyon-proje-maks-merkezicerik Accessed March 29, 2020.
17 https://www.tobb.org.tr/BilgiErisimMudurlugu/Sayfalar/Eng/KurulanKap-ananSirketistatistikleri.php Accessed March 30, 2020.
18 https://www.tcmb.gov.tr/wps/wcm/connect/b4628fa9-11a7-4426-aee6-dae67fc56200/KFE-Metaveri.pdf?MOD=AJPERES&CACHEID=ROOT-WORKSPACE-b4628fa9-11a7-4426-aee6-dae67fc56200-mDXEz4N Accessed March 30, 2020.
19 https://data.tuik.gov.tr/Search/Search?text=HBS Accessed May 12, 2021.
20 https://tuikweb.tuik.gov.tr/PreIstatistikMeta.do?istab_id=334 Accessed March 5, 2021.

References

Balamir M (1985) *Konut Kesimi Araştırmalarında Bilgi Kaynakları ve DİE 1985 Konut Kira Sayımı*. Unpublished report presented at TURKSTAT meeting.
Ball M, Harloe H, Martens M (1988) *Housing and Social Change in Europe and the USA*. London and New York: Routledge.
Ge XJ, Hartfield T (2007) *The Quality of Data and Data Availability for Property Research*. Pacific Rim Real Estate Society Conference Proceeding The 13th Pacific Rim Real Estate Society Conference, 1–12. Available at: https://opus.lib.uts.edu.au/ handle/10453/7375. Accessed April 5, 2020.
Igan D, Loungani P (2012) *Global Housing Cycles*. IMF Working Paper Research Department. Available at: https://papers.ssrn.com/sol3/papers.cfm?abstract_id=2169761. Accessed April 5, 2020.
Kaya A, Topaloğlu Bozkurt A, Bastan EM et al. (2013) *Constructing a house price index for Turkey*. IFC Bulletin no:36. Available at: https://www.bis.org/ifc/publ/ ifcb36.html. Accessed April 5, 2020

Ministry of Development (2018) *11. Kalkınma Planı, Konut Politikaları Özel İhtisas Komisyonu Raporu*. Publication No: 3016-ÖİK:797, Ankara.

Özdemir Sarı ÖB (2019) Redefining the Housing Challenges in Turkey: An Urban Planning Perspective. In: Özdemir Sarı ÖB, Özdemir SS and Uzun N (eds) *Urban and Regional Planning in Turkey*. Switzerland: Springer, 167–184.

Sarıoğlu GP (2007) Türkiye'de Konut Araştırmaları için Veri Kaynakları ve Geliştirme Olanakları Üzerine. *Planlama* 2: 43–48.

Sarıoğlu Erdoğdu GP (2011) Problems in housing research and comparative housing studies. *Megaron* 6(2): 71–78.

Sarıoğlu Erdoğdu GP (2013) *Entry to home ownership: A comparison between Turkey and The Netherlands*. Published PhD Dissertation. Saarbrücken: Lambert Academic Publishing.

TURKSTAT (2015) News Bulletin No: 18631, November 2015, Ankara.

TURKSTAT (2017) *Official Statistics Programme* (Resmi İstatistik Programı) 2017–2021. Publication No: 4443, Ankara.

TURKSTAT (2020a) News Bulletin No: 33820, September 2020, Ankara.

TURKSTAT (2020b) News Bulletin No: 33885, November 2020, Ankara.

TURKSTAT (2020c) News Bulletin No: 33782, November 2020, Ankara.

Van Schie ECM (2006) Knowledge-based housing policy: The Dutch example. Paper presented at the ENHR conference "Housing in an expanding Europe: Theory, policy, participation and implementation", Ljubljana, Slovenia, July 2–5, 2006.

Weller SC, Romney AK (1988) *Systematic Data Collection*. Qualitative Research Methods Series 10, Newbury Park: SAGE Publications.

Part II
Housing, Economy, and Financialization

6 Housing Market Dynamics

Economic Climate and Its Effect on Turkish Housing

Leyla Alkan Gökler

Introduction

Ankara was assigned as the capital city of Turkey after the proclamation of the Republic in 1923 when new housing policies were focussed mainly on resolving the housing shortage in the city during the early urbanization process. After the rapid urbanization of the 1960s, mass migration from the rural to urban areas was witnessed, and squatter settlements started to emerge, and subsequent housing policies were aimed primarily at resolving the squatter problem. Successive amnesty laws served to legalize the squatter settlements, encouraging further illegal housing development (Özdemir, 2010), with further policies developed to steer the transformation of squatter settlements, along with urban transformation projects. These projects, however, led to an increase in rents in the residential areas, leading to uneven profit distribution across the urban space. In Turkey, housing policies are aimed at resolving the consequences of the problems rather than getting to their root. The attempts to resolve the housing problems of the low- and moderate-income groups have included regulation of the housing credit market, mass housing projects aiming to attract customers with lower interest rates, etc.; however, these projects have also made little impact on the problem. The private sector has long been the main actor in the provision of housing in Turkey, as the state did not become a direct provider of housing until 2000 (Özdemir, 2010). As such, the housing market is directly affected by the economic conditions in the country.

The main focus of housing policies since the proclamation of the Republic has been increasing the homeownership rate, and new housing developments are always promoted. While there is a continuous supply of housing to the market, however, the purchasing power of the low- and moderate-income groups is constantly decreasing due to the unstable economic environment in the country. Turkey experienced a couple of deep and prolonged economic crises in the 2000s, and the Turkish housing market has been shaped under the influence of this political and economic instability. As a result of fluctuations in the economy, household incomes are rising slower than home prices, especially for the lower- and middle-income groups, and this

DOI: 10.4324/9781003173670-8

deepened the housing affordability problems witnessed after the 2000s. According to Turkish Statistical Institute (TURKSTAT) data, expenditures on housing and rent are increasing in household budgets, and no permanent policies can be developed to resolve the housing problems of the low- and middle-income groups.

Just as the economy affects the country's housing market, the housing market is highly influential on the economic environment of the country. In Turkey, the construction sector is considered the locomotive of economic growth due to its strong backward and forward linkages with other sectors (Balaban, 2012). Accordingly, the housing market plays an important role in Turkey's economy, and the state always promotes new housing developments to stimulate the economy. Steady growth in total construction output was witnessed up until the 2000s when Turkey faced some economic crises (Berk and Biçen, 2018). In the post-2001 period in Turkey, the housing construction and credit markets started to suffer from the economic crises (Erol, 2019), although in the years that followed, the Turkish construction sector maintained a trend of overall growth, despite some fluctuations (Berk and Biçen, 2018).

Housing production at this time was not developed within the framework of a particular plan in Turkey. As the main driver of the economy, new housing constructions ensure the continuity of the system. In order to develop effective housing policies, both the supply and demand sides need to be taken into account; however, in Turkey, this continuous housing production is more supply oriented and less focussed on the demand side created, especially by the low- and middle-income groups. As the supply side dominates the housing market in Turkey, and the housing sector is affected by most of the upheavals in the political and economic environment, it is difficult to reach an equilibrium in housing supply and demand.

As a result of the crisis experienced in Turkey in recent years, both the demand and supply side of the housing has been affected. Usually, supply creates its own demand, but the decline in purchasing power has had a marked effect on housing sales, while the ongoing supply of housing to the market causing surplus housing stock. Different policies are being introduced to sell these new dwelling units to ensure the economic cycle, such as those encouraging property sales by slashing mortgage interest rates. However, such policies offer only temporary solutions and fail to address the core problems in the Turkish housing market.

Facilitating the acquisition of a real estate by foreigners is seen as a means of stimulating the housing market and financing the rising current account deficit in Turkey in the wake of the recent economic crises. That said, the issue of real estate sales to foreign investors in Turkey has long been an issue of intense debate, and there have been frequent legislative changes. The acquisition of a real estate by foreigners was regulated under Land Registry Law (Law No. 2644) in 1937, and up until 2012, many amendments were made to the Law. These amendments resulted in a significant increase in the sale of houses to foreigners after the 2000s, with the most remarkable take-off being seen in 2018. This new demand by foreigners also affects the housing market and the type of housing projects conducted in the country, spurring an increase in the construction of luxury housing developments in some Turkish cities, and many developers are focussing on this specific market. This new demand also leads to an increase in local housing prices, as stated in several academic studies (Chao and Yu, 2015; Wong, 2017; Liao et al., 2015), in which it is shown that an increase in demand among foreign buyers has the potential to deepen the affordability problems of local buyers.

To sum up, the housing market in Turkey is directly affected by the economic environment in the country and by the different economic policies of the state. Unfortunately, these misguided policies place significant pressure on the housing sector, meaning that policy evaluation is essential for the improvement of the housing market. All of these problems discussed above combine to cause a housing market bubble in Turkey, and the housing policies applied to date have fallen a long way short of being able to resolve the problem. Permanent solutions are needed rather than temporary remedies, and recognizing this, the intention in the present study is to investigate the dynamics of the Turkish housing sector and the effect of the economic pressures seen in the 2000s. To this end, after examining the general status of the Turkish economy in the 2000s, this study analyzes the effects of the economic environment on the housing sector and housing policies in Turkey by focussing on the different problem areas discussed above, drawing upon key sources of data related to the Turkish economy and housing the market, being TURKSTAT and the Central Bank of the Republic of Turkey (CBRT).

This chapter is organized as follows. The following section provides some background information about the economic environment in Turkey after the 2000s. The next section makes a summary of the housing market dynamics in Turkey, followed by a discussion section, while the final section concludes the study.

Economic Environment in Turkey after the 2000s

Turkey has experienced several short-term economic crises in recent decades, the most notable of which were of 1994, November 2000, and February 2001 (Şenses, 2003), which were followed by the 2008 and 2018 crises. Due to the unstable economic environment and the country's weak institutional structure, the Turkish capital markets have been characterized by considerable uncertainty for decades, up until the early 2000s (Erol, 2019).

The structure and the consequences of these economic crises differ from each other. After the 2001 crisis, growth in household debt was witnessed, and real interest rates remained high during this period (Akcay and Güngen, 2019). The crises in 2000 and 2001 stemmed from Turkey's own structural and financial constraints at a time when it was possible to find international financial sources and assistance for the management of both crises (Temiz and Gökmen, 2010). However, the Global Financial Crisis of 2008 that started in the United States was somewhat different, extending its effects into many other countries, spurring record unemployment and severe drops in GDP and industrial output (Rodrik, 2012). The negative effects on Turkey, however, were limited (Coşkun, 2016) due to the recently developed mortgage finance system, the absence of a secondary mortgage market, the less diversified mortgage finance products, the highly regulated financial sector, and lower reliance on mortgage financing in housing transactions (Özdemir Sarı, 2019). In the wake of the 2008 crisis, however, household debt peaked in 2013 in Turkey due to the increasing interest rates associated with the adverse international financial conditions, and in this period, the depreciation of the TRY served to push inflation even higher (Akcay and Güngen, 2019). August 2018 marked the beginning of an exchange rate crisis, in which the rapidly depreciating TRY brought many companies with foreign currency debts to the edge of bankruptcy (Çakmaklı et al., 2020). The TRY lost 31 percent of its value against the US Dollar in 2018 (Akcay and Güngen, 2019) after the crisis, and while still struggling under the effects of the 2018 crisis, the Turkish economy had to deal also with the global COVID-19 pandemic that emerged in Wuhan City in China and quickly spread around the globe. After the 2018 crisis, the TRY tumbled against the dollar, placing a further burden on the cash-strapped Turkish economy, which had already been ailing prior to the emergence of the COVID-19 pandemic (Taskinsoy, 2020).

The TRY continued its downward slide against the US Dollar and Euro after the 2000s. By the beginning of 2020, the TRY had weakened to 6.59 to the Euro and 5.93 to the US Dollar under the economic effects of the global COVID-19 pandemic and hit record lows against the Euro and the US Dollar in December 2020, at 8.02 TRY to the US Dollar, and 9.47 TRY to the Euro, according to CBRT (2020a) data. The foreign exchange market not only affects households and firms who purchase or sell foreign goods or services but also other households, as the depreciation of the currency has a direct effect on inflation in the market. All countries consume foreign products and services, and so imported goods became more expensive with the depreciation of the Turkish currency, leading also to a rise in the price of domestic goods and services. Figure 6.1 presents the consumer price index (CPI) in Turkey as one of the main indicators of inflation.

Figure 6.1 shows one of the effects of the depreciation of the TRY on the economy. This graph of reporting CPI data is based upon a 2003 base of 100 and shows the average yearly change in the price of all goods and services in the economy. A steady and relatively low increase can be seen in Turkey's CPI between 2003 and 2016, and a significant jump in the CPI, especially after 2017, with a 50.98 increase in 2018 when compared to the previous year. The CPI reached 469.59 in 2020, equating to 369.59 percent inflation since 2003.

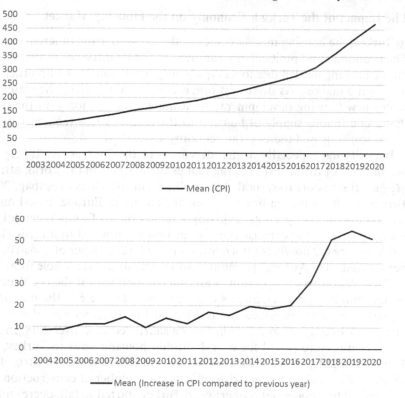

Figure 6.1 Consumer price index (CPI), 2003–2020, (2003 = 100).

Source: Prepared by the author based on TURKSTAT (2020a) data.

In addition to the high inflationary environment in Turkey, unemployment started to increase after 2018, reaching 11 percent in 2018, according to TURKSTAT (2020b) data, up from 9.9 percent in 2014, peaking at 13.7 percent in 2019, and changing little in 2020 at around 13.2 percent. While most global labour markets have been affected by the COVID-19 pandemic, it can be understood from the TURKSTAT (2020b) data that the rise in unemployment is linked to the economic problems before the pandemic.

Turkey has been gone through several financial crises since the 2000s and has implemented many different policies intent on bringing the country out of the crisis that had also affected the housing market. As discussed previously, these crises can be attributed to different factors, although the consequences have been similar, including rising inflation, higher interest rates, TRY depreciation, high unemployment, etc. The following section investigates how the unstable economic and political environment is reflected in the housing market.

The Impact of the Turkish Economy on the Housing Market

In Turkey, the housing market comes under pressure from different factors. The economic and political environment in the country has a direct effect on the housing market due to its input–output relationships with other sectors in the market. As discussed previously, in every crisis, the state promotes new housing developments to stimulate the economy, in the belief that a continuous supply of housing to the market keeps the construction sector working, and thereby, the economy alive.

Intensifying, especially in the aftermath of the 2008 crisis, Turkey has become one of the fastest-growing real estate markets in the world, attracting global investors to several metropolitan Turkish cities (Yeşilbağ, 2019). Turkey has been the leading provider of housing in Europe, based on the issued construction permits, enjoying a large inflow of foreign capital and a sustained housing construction boom that culminated in a crisis (Erol, 2019). Figure 6.2 provides information on the total number of construction permits issued in Ankara, İstanbul, İzmir, and Turkey as a whole from 2015 to 2020. A construction permit is an official approval for the construction of new building projects, and as can be seen in Figure 6.2, the number of construction permits issued for residential buildings witnessed a steady increase between 2015 and 2017 but saw a sharp decrease after 2018. In 2016 over permits were issued for over 1 million housing units in Turkey, and 1,397,778 in 2017, as the most prolific year on record since the turn of the century. With the economic crisis of 2018, the number of construction permits issued for residential properties in Turkey started to fall, decreasing to 323,352—the lowest level since 2003, and similar trends were noted also in Ankara, İstanbul, and İzmir.

Figure 6.2 also provides information about the increase in the number of households when compared to the previous year in Turkey, Ankara, İstanbul, and İzmir between 2015 and 2020, and the need for additional housing is discussed with a comparison of the increase in the number of households and the newly constructed housing units each year. The impact of migration on housing needs cannot be estimated from this graph, being rather a simple comparison of the housing need and the growth in the number of households. Up until 2017, there was an excess supply of housing to the market in Ankara, İstanbul, İzmir, and Turkey as a whole. In 2018, the number of new housing units and the increase in the number of households were closer to each other, aside from in İzmir. For instance, in 2017, the number of newly constructed housing units was almost three times the additional number of households in Turkey. That is, an excess of more than 900,000 housing units were introduced to the market that year, with an excess of 65,000 in Ankara, 187,000 in İstanbul, and 55,000 in İzmir. As discussed earlier, in 2019, the construction of new housing saw a sharp decline, with fewer new properties built than the increase in the number of households. A small increase in the housing construction sector was seen

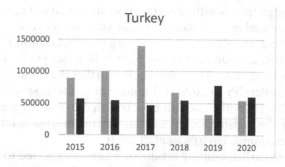

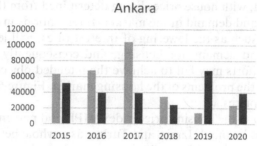

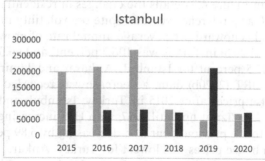

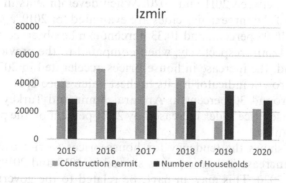

Figure 6.2 The number of new housing units and increase in the number of house-holds between 2015 and 2020.

Source: Prepared by the author based on TURKSTAT (2020c, 2020d) data.

in 2020, although new housing construction struggled to keep pace with the growth in the number of households that year. In short, as can be understood from the data in Figure 6.2, housing construction has been affected by the economic environment of the post-2018 crisis era in Turkey and in three metropolitan cities. The figure shows further that housing production, which fell considerably in 2019, still managed to increase, despite the pandemic, in 2020. This little rise can be attributed to several measures aimed at supporting the housing sector, such as cutting mortgage interest rates in the unstable economic environment of 2020.

As discussed earlier, housing production can be expected to be shaped by housing demand, with house prices being determined from the equilibrium point of supply and demand in the market. Interventions in support of the housing sector, such as the lowering of interest rates, will lead to changes in the supply and demand for housing, and consequently, house prices, although such efforts may fail to achieve the intended objectives, and may actually deepen the problems of the housing market. Figure 6.3 presents the trends in house prices.

Figure 6.3 shows the housing price index (HPI) and percentage change in HPI for Ankara, İstanbul, İzmir, and Turkey as a whole between 2010 and 2020. The HPI in Figure 6.3 details the changes in residential house prices using 2017 = 100 as a reference year. Despite the volatility in house prices, the general trend is upward. The average annual rate of growth in the HPI between 2010 and 2020 in Turkey was 106.7 percent, and 96.3 percent, 94.8 percent, and 117.5 percent for İstanbul, Ankara, and İzmir, respectively. According to CBRT (2020b) data, a decrease was recorded in the rate of increase in house prices after 2015 in Turkey, İstanbul, and Ankara, but an upward trend for İzmir up until 2017, with HPI increasing by 18.63 percent compared to the previous year and decreasing by 5.89 percent in 2018. The increase in house prices was lowest for Turkey, Ankara, İstanbul, and İzmir in 2018 between 2011 and 2020. When developments in the housing price indices of the three major cities are evaluated for 2019, again, a rise of 3.63 percent, 10.08 percent, and 10.33 percent can be observed for İstanbul, Ankara, and İzmir, respectively, when compared to the previous year. On the other hand, the increase in house prices accelerated in 2020, when the house price growth indicator hit its highest value since 2011, recording an increase of around 30 percent in Ankara, İzmir and Turkey as a whole, while the House Price Index increased by 27.91 percent on the previous year in 2019 in İstanbul.

Figure 6.3 shows the tendency for house prices to rise over the years, as while the increase in house prices slowed in 2018 and 2019, it began to accelerate in 2020. This may, in part, be related to the government intervention in the housing sector. As discussed previously, the interest rate on mortgages was cut to the lowest value seen in the 2000s in an attempt to stimulate home purchases, but the resulting increase in housing demand led also to a rise in house prices in 2020, to reach the highest level since 2011.

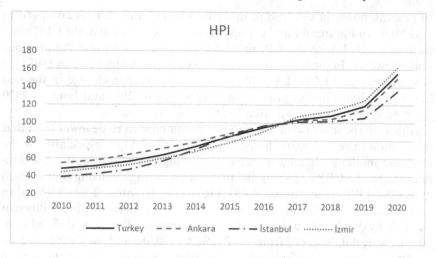

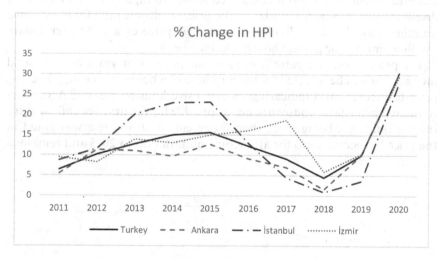

Figure 6.3 Housing price index (2017 = 100).

Source: Prepared by the author based on CBRT (2020b).

Another reason for this rise may be related to the increasing demand for detached houses after the emergence of the COVID-19 pandemic in 2020, with increased interest being witnessed in properties with gardens or balconies in the suburbs. The result has been an increase in sale and rental prices of detached houses (Zoğal and Emekli, 2020; Şarkaya İçellioğlu and Güler, 2020; Şükürlüoğlu, 2020).

Households in Turkey invest in housing as a means of protecting themselves against the volatile economy, with the expectation that the increase in house prices will exceed that of inflation (Yıldırım and İvrendi, 2017).

In general, house prices tend to move in line with inflation. A comparison of the data in Figures 6.1 and 6.3 reveals a significant jump in the CPI, especially after 2017, with the CPI in 2018 increasing by 50.98 on the previous year. However, Figure 6.3 indicates a slowdown in the increase in HPI that year. According to CBRT (2020b) data, the percentage change in the real house price index had negative values between July 2017 and January 2020 but moved into the positive after 2020, as discussed above.

Although the increase in housing prices can be seen to be lower than that of inflation in recent years, housing prices are still high, especially for the low- and moderate-income groups. According to housing unit price data released by the CBRT (2020c), the price of an average housing unit in 2020 was 3,888 TRY/m² in Turkey, 6,483 TRY/m² in İstanbul, 2,895 TRY/m² in Ankara, and 4,505 TRY/m² in İzmir. A simple assessment of affordability can be made by dividing house prices by annual income (Rodda and Goodman, 2005). Using data sourced from TURKSTAT (2020e) and the CBRT (2020c), the Price to Income Ratio is calculated for an average of 100-meter square dwelling units. Figure 6.4 illustrates the price to income ratio for five income quintiles, calculated by dividing the average price of a 100-meter square dwelling unit by the annual household income.

The price-to-income ratio illustrates the number of years of household income that will be invested for the purchase of a home, with a high housing price-to-income ratio indicating an expensive housing market. According to the available data, houses are not affordable for the first, second, or third income quintiles, being more than 6 percent. A steady rise was noted in the price to income ratio for all income quintiles up to 2015 and remained

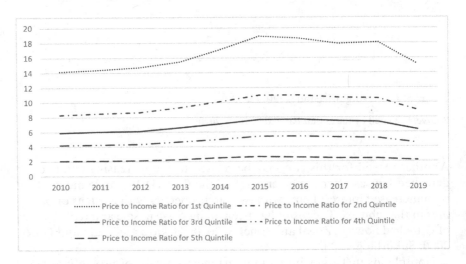

Figure 6.4 Price to income ratio.

Source: Prepared by the author based on CBRT (2020c) and TURKSTAT (2020e) data.

high until 2018. The price to income ratio seems to be more stable for the fifth income quintile, with values of between 2.05 and 2.62 in the related period. However, the price to income ratio of the first quintile shows more volatile values, changing from 14.15 to 18.88. These figures show how economic problems affect different income quintiles. A significant gap between the first income quintile and the other quintiles can also be seen in the figure, revealing an important housing problem for the first quintile related to affordability—that the first three income groups in Turkey are struggling with high house prices and low incomes. Furthermore, according to TURKSTAT (2019) data on household consumption expenditures in 2019, the largest share of household budgets is allocated to housing and rent, being around 24 percent in Turkey.

Figure 6.4 has shown that potential home buyers face high house prices, while CBRT (2020b) data shows that percent change in the real house price index had a negative value between July 2017 and January 2020, creating difficulties for both home producers and home sellers. That is, on the one hand, house prices are quite high when compared to household income, while on the other hand, the increase in house prices was below the general inflation rate between 2017 and 2020. In order to discuss this dilemma, it will be beneficial to look also at house sale numbers.

As can be seen from Figure 6.5, housing sales followed a rising trend in Turkey between 2013 and 2017 and then started to decrease in 2018 and 2019 but reached their highest level since 2013 in 2020. A similar trend can be noted for Ankara and İzmir, with a decreasing trend in 2018 and 2019, followed by a rise in 2020, while the figures for İstanbul are somewhat steadier.

The statistics presented in Figures 6.3, 6.4, and 6.5 indicate that the housing market in Turkey is significantly affected by economic conditions in the country. There has been a significant decrease in the number of new housing units after 2017, and house sales also started to decrease, with the percentage of change in the real house price index having negative values after 2017. Despite the negative indicators in the housing market, more favourable conditions began to emerge in 2020. As discussed previously, in every crisis, the state promotes new housing developments in order to stimulate the economy, and in 2020, developed new policies related to the housing market for the same purpose, one of which was related to mortgage interest rates.

The current mortgage law in Turkey was enacted in 2007, prior to which commercial banks would provide mortgage credits with high-interest rates of between 1.52 and 4.02 in the 2003–2007 period, according to CBRT (2020d) data. After the law entered into force, mortgage interest rates started to fall, declining to 0.81, which is still high for the low- and moderate-income groups. So despite the relative decline in interest rates, the current Mortgage Law still falls short of addressing the deficiency in the housing market in Turkey, given that housing credits remain out of reach for the low- to moderate-income

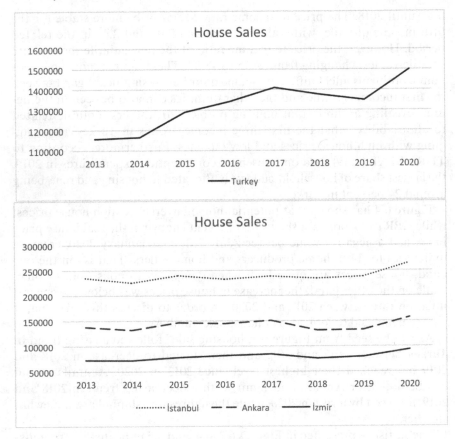

Figure 6.5 House sales by provinces and years, 2013–2020.

Source: Prepared by the author based on TURKSTAT (2020f) data.

groups (Alkan, 2015). After the 2018 crisis, interest rates rose to 1.61, again putting housing credits out of reach of most income groups.

As the figures illustrate, the overall performance of the housing market has been unsatisfactory since 2017, and the pandemic in 2020 added to the woes of the economy and the housing market. In response, public banks cut mortgage interest rates to 0.64 percent for new-build sales, and 0.74 for second-hand sales, with a grace period of up to 12 months, in order to stimulate the economy. This led to a jump in property sales across Turkey to 190,012 in June and to 229,357 in July—a clear indication of the success of the policies enacted to revive the housing sector in increasing housing sales.

Foreign real estate investments have also been seen as a means of financing the rising current account deficit in Turkey following the economic crises. Foreign direct investments (FDI) have led the real estate markets to grow since the 2000–2001 crisis in Turkey, although the acquisition of a

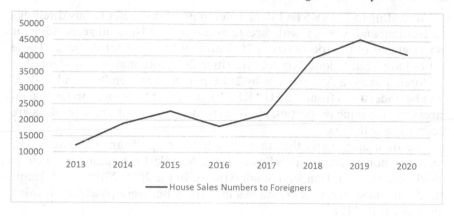

Figure 6.6 House sales numbers to foreigners, 2013–2020.

Source: Prepared by the author based on TURKSTAT (2020g) data.

real estate by foreign investors in Turkey has long been an issue of intense debate, and there have been frequent legislative changes. The acquisition of a real estate by foreigners was regulated previously by the Land Registry Law (Law No. 2644), which dates back to 1937, but between 1937 and 2012, many amendments were made to the Law, and arrangements to Land Registry Law after the 2000s have eased the acquisition of a real estate by foreigners.

Figure 6.6 provides information on the number of house sales made to foreigners in the 2013–2020 period, showing a rising trend over the years and a marked increase, particularly in 2018. As discussed previously, there has been a continuous decline in the value of the TRY against the US Dollar as a result of financial crises. Such currency devaluations in a country usually encourage foreign purchases of real estate, as foreign buyers can buy large amounts of attractively priced property. It comes as no surprise, therefore, that in 2018 foreign home purchases rose by almost 78.4 percent on the previous year to 39,663 dwelling units, and there was strong growth also in 2019 when the total number of foreign sales reached 45,483 units based on TURKSTAT statistics. This high demand for foreign property in Turkey in recent years may be due mainly to tourism and the devaluation of TRY. As expected, the number of foreign home purchases declined in the first half of 2020 due to the COVID-19 pandemic, but if the devaluation of the TRY continues, foreign demand will again be encouraged next year, and the increase in foreign buyers may push real estate prices up to levels beyond the affordability of the residents. It was shown earlier in Figure 6.3 that house prices tend to rise in Turkey from year to year, and that in general, the house price to income ratio is very high in the country, which is unsustainable in the long term. In addition to these problems, an increase in foreign house buyers may lead to local buyers being priced out of the market, for whom prices are already at unsustainable levels.

According to TURKSTAT data, foreign investors prefer to buy dwellings in İstanbul and Antalya, with foreign home sales in İstanbul accounting for 45.9 percent of the total sales in 2019 and 19.7 percent for Antalya. These cities have been followed by Ankara since 2019, although the total number of foreign sales in Ankara rose by 210.8 percent between 2017 and 2019. It can be understood from the TURKSTAT data that foreigners prefer coastal areas when buying properties, as all cities other than Ankara in the data are located near the sea.

The data also shows that there has been a significant increase in the share of house sales to purchasers from Middle East countries, such as Afghanistan, Iraq, Iran, and Saudi Arabia. In the 2015–2019 period, Iraqis made the most purchases of homes in Turkey but were passed by Iranian purchasers in 2020.

It is assumed that the Turkish real estate market will continue to be attractive for foreigners looking for high investment opportunities at affordable prices and Golden Visas for residency permits if TRY continues to depreciate. In 2019, 1,348,719 dwelling units were sold, of which 45,483 were sold to foreigners, corresponding to around 3.38 percent of total sales. In short, foreign buyers are purchasing a large number of homes in the Turkish housing market that many local buyers cannot afford.

Discussions

The Turkish economy has faced several economic crises since the turn of the century, leading to a rapid depreciation of the TRY, high unemployment, and rising inflation. Economic indicators produced by TURKSTAT and CBRT reveal significant problems related to the Turkish economy, especially after the 2018 crisis, after which a significant jump was seen in the CPI, and within this high inflationary environment, unemployment started to increase also after 2018. Furthermore, the TRY has hit record lows against the Euro and the US Dollar after 2020.

The housing sector has been significantly affected by the economic environment in the country since 2018. According to available data, construction permits saw a significant decrease after 2018, while in 2019, the construction of new housing units left behind the increase in the number of household size. According to CBRT (2020b) data, the increase in house prices reached its lowest in 2018 since 2011, and the percent change in the real house price index had negative values between July 2017 and January 2020, and house sales also started to decrease after 2018.

The statistics used in this manuscript reveal the housing market in Turkey to be significantly affected by the economic environment of the country, especially after 2018, and it can be argued that there is also a crisis in the housing market, especially after 2018, which creates problems for producers and consumers alike. The producers experience a loss of profit as the increase in the price of housing lags behind inflation, and due to the rising price of inputs related to the depreciation of the TRY in the related period, and producers may also experience difficulties in selling their products. Economic crises in the housing sector also affect house buyers, as housing prices are still considerably higher than the household income. Considering the increasing inflation and unemployment rates in recent times, buying a home would seem to be quite difficult for many households.

This situation is deepening the housing problem and is also making it difficult to establish a balance between supply and demand in housing. The successive economic crises serve to increase the accessibility problems of the low- and middle-income groups, leading housebuilders to produce to meet the demand of the upper-income groups as a more viable option. Thus, the construction sector, which is constantly striving to survive, is moving further away from solutions that will satisfy the demands of the lower-income groups, leading also to the exclusion of low- and moderate-income families from the housing market, who already have limited access to housing in Turkey.

The data also revealed that most of the indicators related to the housing market started to improve after 2020, despite the continued economic instability brought by the COVID-19 pandemic. This can be attributed to the government interventions into the housing market in 2020 aimed at stimulating the economy, although these positive indicators will not be sustainable if economic conditions do not improve in the coming years.

Policymakers have enacted other policies in an attempt to stimulate the housing market, and foreign investment into real estate is being considered as a partial solution. Since the early 2000s, there has been a significant increase in the sale of housing to foreigners after a remarkable upturn in 2018. This new demand among foreigners is also affecting the housing market and the types of housing projects in the country, tending to increase the construction of luxury housing developments in some Turkish cities. Many developers have started to focus on the foreign housing demand, which leads also to an increase in local housing prices. As discussed previously, there have been studies supporting these arguments, warning that the increased demands of foreign home buyers from the Turkish housing market may deepen the affordability issues faced by local buyers. In 2019, sales of housing to foreigners amounted to around 3.38 percent of the total. The purchasing power of the low- and moderate-income groups is constantly decreasing due to the consecutive economic crises in the 2000s. The high inflation and the continued depreciation of the TRY deepen the housing affordability problems of the low- and middle-income groups, while there is increasing demand among foreign homebuyers who are looking to take advantage of the economic woes in the country.

Conclusion

This chapter has examined the link between the Turkish housing market and the economic environment in the country. The assumption that the housing market is being pressured by the economic and political environment is tested based on basic economic indicators within the Turkish economy and an analysis of the housing market since the turn of the century. As can be understood from the data garnered for this study, the economic environment has had a direct effect on housing policies. The housing market is always promoted in the country as a means of keeping the construction sector alive, but unfortunately, the continued housing supply is not being managed according to a specific plan, leading to several physical and social problems, especially in Turkey's metropolitan cities. Turkey has experienced a couple of serious long-term economic crises that have resulted in high inflation, a depreciation of the TRY, and high mortgage interest rates, making housing less affordable. In addition to these problems, the housing market in Turkey is always under considerable pressure, and the housing policies applied to date have fallen well short of resolving the problems and have actually contributed to the creation of further problems.

References

Akcay Ü, Güngen AR (2019) The making of Turkey's 2018–2019 economic crisis. Working Paper, No. 120/2019, *Hochschule für Wirtschaft und Recht Berlin*. Berlin: Institute for International Political Economy (IPE).

Alkan L (2015) Effects of socio-economic factors on home purchases: The cases of Yenimahalle and Çankaya in Ankara. *International Development Planning Review* 37(3): 289–309. DOI: 10.3828/idpr.2015.8.

Balaban O (2012) The negative effects of construction boom on urban planning and environment in Turkey: Unraveling the role of the public sector. *Habitat International* 36(1): 26–35.

Berk N, Biçen S (2018) Causality between the construction sector and GDP growth in emerging countries: The Case of Turkey. *Athens Journal of Mediterranean Studies* 4(1): 19–36.

CBRT (2020a) Foreign Exchange Rates (Dollar and Euro). Available at: https://evds2.tcmb.gov.tr/index.php. Accessed March 17, 2021.

CBRT (2020b) Housing Price Index. Available at: https://evds2.tcmb.gov.tr/index.php. Accessed March 17, 2021.

CBRT (2020c) Housing Unit Price. Available at: https://evds2.tcmb.gov.tr/index.php. Accessed March 17, 2021.

CBRT (2020d) Mortgage Interest Rate. Available at: https://evds2.tcmb.gov.tr/index.php. Accessed March 17, 2021.

Chao CC, Yu ESH (2015) Housing markets with foreign buyers. *The Journal of Real Estate Finanance and Economics* 50: 207–218. DOI: 10.1007/s11146-014-9454-3.

Coşkun Y (2016) Housing finance in Turkey over the last 25 years: Good, bad or ugly? In: Lunde J, Whitehead CME (eds) *Milestones in European Housing Finance*. United Kingdom: Wiley, 393–411.

Çakmaklı C, Demiralp S, Kalemli Özcan Ş et al. (2020) COVID-19 and emerging markets: The case of Turkey. *Koç University-Tüsiad Economic Research Forum Working Paper Series*. Available at: https://ideas.repec.org/p/koc/wpaper/2011.html. Accessed June 16, 2021.

Erol I (2019) New geographies of residential capitalism: Financialization of the Turkish housing market since the early 2000. *International Journal of Urban and Regional Research* 43(4): 724–740.

Liao WC, Zhao D, Lim LP et al. (2015). Foreign liquidity to real estate market: Ripple effect and housing price dynamics. *Urban Studies* 52(1): 138–158.

Özdemir D (2010) The role of the public sector in the provision of housing supply in Turkey, 1950–2009. *International Journal of Urban and Regional Research* 36(6): 1099–1117.

Özdemir Sarı ÖB (2019) Redefining the housing challenges in Turkey: An urban planning perspective. In: Özdemir Sarı ÖB, Özdemir SS, Uzun N (eds) *Urban and Regional Planning in Turkey*. Switzerland: Springer, 167–184.

Rodda DT, Goodman J (2005) *Recent House Price Trends and Homeownership Affordability*. Prepared for the Office of Policy Development & Research. U.S. Department of Housing and Urban Development, Washington.

Rodrik D (2012) The Turkish economy after the global financial crisis. *Ekonomi-tek* 1(1): 41–61.

Şarkaya İçellioğlu C, Güler MG (2020) Covid-19 pandemisinin emlak piyasası üzerindeki etkileri. 9. Uluslararası Meslek Yüksekokulları Sempozyumu, 18–20 Kasım 2020, İstanbul.

Şenses F (2003) 5 Economic crisis as an instigator of distributional conflict: The Turkish case in 2001. *Turkish Studies* 4(2): 92–119.

Şükürlüoğlu R (2020) Yeni ekonomide gayrimenkul piyasalarındaki güncel gelişmeler ve piyasa analizi. 9. Uluslararası Meslek Yüksekokulları Sempozyumu, 18–20 Kasım 2020, İstanbul.

Taskinsoy J (2020) COVID-19 Could Cause Bigger Cracks in Turkey's Fragile Crisis Prone Economy. Available at: https://papers.ssrn.com/sol3/papers.cfm?abstract_id=3613367. Accessed April 12, 2021.

Temiz D, Gökmen A (2010) The comparative effects of 2000–2001 financial crises and 2008 global crisis on the Turkish economy. *International Journal of Economic and Administrative Studies* 2(4): 1–24.

TURKSTAT (2019) Household Consumption Expenditure. Available at: https://biruni.tuik.gov.tr/medas/?kn=132&locale=tr. Accessed March 17, 2021.

TURKSTAT (2020a) Consumer Price Index. Available at: https://biruni.tuik.gov.tr/medas/?kn=84&locale=tr. Accessed March 17, 2021.

TURKSTAT (2020b) Labor Statistics. Available at: https://data.tuik.gov.tr/Kategori/GetKategori?p =Istihdam,-Issizlik-ve-Ucret-108. Accessed March 17, 2021.

TURKSTAT (2020c) Building Permit Statistics. Available at: https://data.tuik.gov.tr/Kategori/GetKategori?p=insaat-ve-konut-116&dil=1. Accessed March 17, 2021.

TURKSTAT (2020d) Total Household Population, Number of Households and Average Size of Households by Locality. Available at: https://data.tuik.gov.tr/Kategori/GetKategori?p=nufus-ve-demografi-109&dil=1. Accessed March 17, 2021.

TURKSTAT (2020e) Distribution of Annual Household Disposable Income (Average, TRY). Available at: https://data.tuik.gov.tr/Kategori/GetKategori?p=gelir-yasam-tuketim-ve-yoksulluk-107&dil=1. Accessed March 17, 2021.

TURKSTAT (2020f) House Sales Statistics. Available at: https://data.tuik.gov.tr/Kategori/GetKategori?p=insaat-ve-konut-116&dil=1. Accessed March 17, 2021.

TURKSTAT (2020g) House Sales to Foreigners. Available at: https://data.tuik.gov.tr/Kategori/GetKategori?p=insaat-ve-konut-116&dil=1. Accessed March 17, 2021.

Wong A (2017) Transnational real estate in Australia: New Chinese diaspora, media representation and urban transformation in Sydney's Chinatown. *International Journal of Housing Policy* 17(1): 97–119. DOI: 10.1080/14616718.2016.1210938.

Yeşilbağ M (2019) The state-orchestrated financialization of housing in Turkey. *Housing Policy Debate* 30(4): 533–558.

Yıldırım MO, İvrendi M (2017) House prices and the macroeconomic environment in Turkey: The examination of a dynamic relationship. *Economic Annals* LXII(215): 81–110.

Zoğal V, Emekli G (2020) The changing meanings of second homes during Covid-19 Pandemic in Turkey. *International Journal of Geography and Geography Education (IGGE)* 42: 168–181.

7 Varieties of Residential Capitalism in Turkey

State-Led Financialization of Housing

Zeynep Arslan Taç

Introduction

There are several studies on financialization from different disciplines and approaches. However, there is little consensus on the definition of the term. Gotham (2009) defines financialization as a multi-national, controversial, and conflicting process that new financial instruments as real estate emerges in different economic areas. Fernandez and Aalbers (2016: 71) define financialization as "the increasing dominance of financial actors, markets, practices, measurements, and narratives, at various scales, resulting in a structural transformation of economics, firms, states, and households". Financialization can transform specific areas of economies and societies like housing and housing classes. The financialization of real estate and housing takes different forms across countries, regions, cities and can be heterogeneous and path-dependent. Specific historical legacies and dynamics shape financialization differently in different geographies.

Since the 2008 global financial crisis emerged in developed economies, the financialization of housing (FoH) literature is mostly debated Global North. As the spread of financial crises, financial instruments also spread to the Global South. On the Global North and South divide, center-periphery literature mostly focuses on foreign direct investment and trade relations. However, the literature on the FoH claims there is a further relation. Financialization interacts with various social institutions such as international ows, loans, and household debt. The volume and debt of the financial sector differ widely across all political economies; the relations between countries are uneven. Because of this heterogeneity, housing-centered financialization shows uneven and combined characteristics (Fernandez and Aalbers, 2019), which affect households' savings and spending. To better understand the FoH, different types of capitalisms—mainly residential capitalism and national politics, which shape the pattern of FoH, needed to be understood. Residential capitalism studies question why and how real estate became an object of financialization. In most economies, residential capitalism is about residential property

DOI: 10.4324/9781003173670-9

ownership and its associated mortgage debt (Schwartz and Seabrooke, 2008). The literature of varieties of residential capitalism largely rests on the evidence from the advanced capitalist, high-income economies parallel to the general trend in financialization studies (Becker, 2013). Less developed, low-to middle-income countries mostly do not capture in Varieties of Capitalism (VoC) literature.

Residential capitalism is diverse as financialization and FoH. Residential capitalism is related to the construction sector because the housing sector is one of the sub-branches of the construction sector. Schwartz and Seabrooke (2008) state that no individual country presents a pure form of residential capitalism. The FoH develops in different ways in Global North and Global South; indicators of housing financialization such as private debt trends and economic typologies show no single model of residential capitalism in the Global South. Fernandez and Aalbers (2019) underline residential capitalism's common denominators in Global South as a combination of relatively low mortgage debt, high ownership rates, a large informal sector with low housing quality, and significant capital inflows (see Ergüven, 2020; Ashman et al., 2011; Fernandez and Aalbers, 2019; Socoloff, 2020; Mawdsley, 2016; Bonizzi et al., 2020). Historical legacies, local dynamics, and high informality harden to follow residential capitalism and FoH in Global South as a linear line compared to Global North. As a dependent state-led market economy, Turkey shows RC signs with relatively high and fixed ownership rates (~60%), low mortgage debt to GDP ratio,[1] and Housing stock, which is completely or partially against Development Law in the total stock, is over 50% (TMMOB, 2004). As a late-capitalist country, the mortgage system is relatively new (established in 2007) and not developed in Turkey. High-interest rates and short-term loans characterize the Turkish mortgage system. With these features, the Turkish mortgage system differs from other countries. On the other hand, informal housing refers to any form of housing that is illegal, including squatter housing and any type of unauthorized housing built without complying with the rules introduced by Development Law, regulations, plans, licenses, and their annexes. Turkey's position as a semi-peripheral country in the global economy, historical legacies and local dynamics as informal housing and clientelist relations fed by the housing sector and transformation of financial architecture after the 1980s, and State's changing role from the regulator to an active market actor are the specific dynamics that shape financialization process.

The aim of this paper is to discuss the role of the housing sector in Turkey's financialization process and explore the informal sector's role. The chapter argues the central role of the State concerning Turkey's financial architecture. The purpose here is to highlight distinctive dimensions of the Turkish case and show how residential capitalism is making progress. Developing a comparative theoretical framework for residential capitalism in Global South is beyond the scope of this study. In this regard, the chapter highlights the historical legacy of informal housing and concludes with an overall evaluation of the national political economy of housing.

Legacy of Informal Housing

The housing regime is a product of a complex historical and political process. It is related to the housing market and reflects the relation between institutions, traditions, and other social dynamics (Bengs, 2015). These dynamics include the structure coalitions, composition, and bargaining capacity of local actors and the family's role in sustaining or supporting entry into homeownership. Limited attention has been paid to the informal housing, informal, and family practices that support entry to homeownership (Allen et al., 2004), whereas there is a well-developed literature on squatter housing and informal settlements is in Turkey. The existence of informal modes of housing provision outside the banks, financial institutions, and planning regulations has been an important feature of urbanization in peripheral countries (Aalbers et al., 2020).

Turkey's case demonstrates the preexisting relationship between the informal housing market and financialization and how the State promotes financialized housing provision. The 1950s has a crucial impact on the Turkish housing regime. Historically, the transition to a multi-party system from the one-party regime and intense migratory flows from rural to urban overlaps. The land rent emerged because the squatter housing sector in cities provided a new capital accumulation tool during this period but started to be realized after the 1980s. With the increasing infrastructure needs of urban Turkey, the gross output of the construction industry—not only for its own sake but also on related goods and services it can create or help to create in the economy discovered in this period.

Furthermore, the State created a clientelist construction elite, as a new capital class, by transferring public resources to mass infrastructure projects, which caused the construction economy based on valuable land rent and infrastructure projects. Especially after the 2000s, massive infrastructure projects and construction economies have become the pillars of the Turkish economy. Social networks continue to exist in informal settlements; the surplus value of land has also been shared informally in the Turkish case. The dynamics of the housing regime and coalition of local actors as capital and rentier classes vary in Turkey's provinces (see Aksu Çam, 2020; Arslanalp, 2018; Danış and Kuyucu, 2020). As an outcome of a less-regulated housing market, the housing starts have a broad geographical variation (Türel and Koç, 2015), which leads to supply–demand imbalances on a regional basis. Due to the variation of local actors and regional differences, the variegated and uneven character of FoH and residential capitalism can be observed at the national level.

Informal settlements and unauthorized buildings vary in form. The most common and known form of unauthorized buildings is squatter housing.

Informal settlements may not have title deeds, built without construction and occupancy permits, or do not have occupancy permits only. In some cases, the buildings with construction and occupancy permits become against construction projects or development legislations due to later additions, such as turning a terrace into a room. There are also unauthorized power plants, piers, shipyards, harbors, and non-building structures such as retaining walls, fill areas, pools, and sports fields. These buildings are not unique to low-income housing areas or built only in the past. 16:9 skyscrapers[2]—located in one of the most valuable regions of İstanbul—with 40% foreign property owners is an example of unauthorized megaprojects[3] built in the last decade. It is estimated that total unauthorized building stock, including informal housing and the apartment blocks built against Development Law before 2018, is over 50% of the total buildings, which corresponds to be around 13 million units (Ministry of Environment and Urbanisation, 2018: 4) while half of these buildings are apartments, the Ministry forced construction firms and contractors to obtain occupancy permits via this regulation. Based on the figures in 2008, the estimated rates for unauthorized buildings are 70%, 30–40%, and 60% for İstanbul, Ankara, and Izmir[4] (Turkish Court of Accounts, 2008: 40). The Ministry of Environment and Urbanisation declared development peace in 2018.[5] The peace aims to legitimize unauthorized building stock and dispose of controversial situations obstructive to urban transformation projects. By the end of 2019, almost 7.5 million independent units have benefited from the redevelopment peace so far and paid approximately 25 billion TL (Şensoy Boztepe, 2020). Cities with the highest number of applications are also the cities with the highest unauthorized building[6] stock (Gündoğmuş, 2019). The Ministry records the revenues in the budget to use in urban transformation projects and provides zero-interest loans to municipalities if they prefer 100% domestic construction materials in urban transformation projects.

Historical Legacies and Local Dynamics

Informal housing is one of the main characteristics of the Turkish housing regime. Although there were practices such as planning the capital and housing provision for civil servants and parliament members, there was no national housing policy from the Early Republican era to the 1950s. The main priority was strengthening Turkey's political and economic independence—there was not enough budget for housing provision. Due to intense migratory flows from rural to urban, massive urbanization brought new problems like the excessive need for housing and a visible shortage of housing. Self-helped, self-built, informal housing on illegally claimed or bought land, which belongs to public or third parties in the peripheries, became the leading solution of migrants who needed shelter. Starting from these years when the housing shortage emerged, the shortage of authorized, decent-quality housing has become a chronic housing system problem. This problem has mostly been addressed as a temporal quantitative deficiency crisis (Özdemir Sarı, 2019) possible to be solved by individual measures, which are ease of access to housing finance and expansion of homeownership.

In the 1950s, the construction and real estate sector's importance for economic growth was recognized for the first time. There was a lack of housing alternatives both in the rental and for-sale housing supply. Land ownership was the first condition of being a homeowner. Before the Flat Ownership Law (1965), individual ownership of a flat was not possible legally, and housing problems tried to be solved through a regulation enabling shared ownership of the land. Although it was limited, individual construction continued in this period. Besides, apartment buildings could be built with informal agreements between the landowner, the contractor, and several households (ownership fragmentation during the construction process and after completing construction was two separate economic processes). Families who do not own land could enter into homeownership via these informal agreements. With the legal recognition of ownership fragmentation in 1965, more than half of the total planned housing stock in Ankara in 1970 seems to have been developed in this way (Balamir, 1975). Although there is no legal document showing the Turkish housing policy is against public or social rental housing, it has never been a policy issue. A limited number of public housing were provided across the country only for certain public officials. Within the housing regime, households have always been encouraged to homeownership. Squatter housing cleared the path of owner-occupancy for the disadvantaged. Although many legal regulations have been enacted to prevent self-helped, self-built squatter housing provision since the 1960s, governments have not strictly enforced the laws for political clientelism. People did not hesitate to resort to illegal means to meet their housing needs, and squatter housing became the norm in urban Turkey until the 2000s. Urban services like transportation provided in these areas as albeit limited compared to other parts of the city. Most of the infrastructure services such as electricity, water, and natural gas provided very late and were only accessible informally at first. These areas did not remain as poor housing complexes. They were commodified and commercialized in time by legalization/formalization. Squatter houses were typically built to be used by the owner; however, signs of commercialization were already present; some of them rented or built to be rented. Not all of the squatters occupied the land in the first place. The land was already trading illegally by non-owners without titles. In the absence of healthy, adequate, and affordable housing, squatter housing was the primary tool to solve the working class's shelter and dwelling problem, which also was a tool to keep the wages at low levels (Tekeli, 2009). Several amnesties granted to squatter housing led to a change in the character of informal settlements that lost their single-story, detached look and turned into five-story apartment blocks. Squatter housing was a source of income for long-standing households, formerly "owner"-occupied neighborhoods became a mix of rental and owner-occupied dwellings (Kızıldağ Özdemirli, 2018), and there is a positive transfer in the income distribution in favor of informal housing owners. (Başlevent and Dayioğlu, 2005).

From Commodification to Financialization: The Role of Informal Housing

The restructuring of the Turkish economy after the 2001 crisis has affected urbanization and housing policy. Financial liberalization and profession-alization of the housing market progressed together. In the 2000s, land occupation and self-helped, self-built informal housing was no more toler-ated (Keyder, 2005). On the other hand, unauthorized buildings could not be prevented by municipalities. While factories moved from city centers to outside, squatter settlements have become the areas where potential land rents are the highest. From the financial perspective, low-income hous-ing areas are assets. *Squatter housing and* unauthorized apartment blocks are not just housing; they are a field of assets waiting to be transformed to capitalize on the potential land rent. State-led urban transformation of squatter settlements starting in the 2000s enabled finance-led housing provision. Urban transformation projects became the primary policy tool for the destruction and reconstruction of existing housing to capitalize on the potential rent. The integration of the poor and low-income families from informal financial services to a formal financial base is key to joining the global economy system (Güngen, 2017). Besides, changes in the tenure structure of squatter housing are not solely through urban transforma-tion projects; informal sales, not integrated into the financial system, are widespread.

Large-scale urban transformation processes started with transforming squatter housing and financed by international credit institutions, related to increasing and reorganizing urban rent (Kayasü and Uzun, 2009: 152). On the other hand, urban transformation projects are not limited to the renewal of informal settlements; new residential areas are also developed under urban transformation projects. The State provides legislative reforms and favorable policies, privatizes public land, and mobilizes large amounts of public funds and investments. It should be underlined that *Housing Development Administration (TOKİ)* is at the heart of the State's heavy involvement as the strongest actor in the market. Via dozens of legislative changes, TOKİ took control of a massive land stock and gained the power to implement development plans. There is an incomparable difference between the lands that TOKİ and contractors reach as sources. The FoH transforms as capital/money travels across the globe. There is an ongoing process in varying forms with similar aims and results. It varies not only in big cit-ies but also observed in small cities. For example, TOKİ is very active not only in metropolitan cities but also in small Anatolian cities. In 2018 more than three thousand dwelling units, mosques, a shopping center, a school, a health center, and a police station constructed by TOKİ in Nevşehir as urban transformation projects (TOKİ, 2018).

Since the 2000s, UTPs have mainly exercised deconstruction and reconstruction of housing to capitalize on the potential rent. Despite the right macroeconomic conditions and the government's will, the *regeneration* policy generated legal uncertainties, institutional clashes, which raised transaction costs and led to failures; only two projects were completed in more than 30 regeneration areas in İstanbul between 2004 and 2019 (Kuyucu, 2020a). The rest never reached the implementation stage or stopped by courts. Urban transformation projects in informal settlement areas serve mainly three aims. Firstly, land, one of the most expensive inputs of the construction cost (Tunç, 2020a: 41), is provided free from squatter owners in exchange for flats.[7] Secondly, informal, unauthorized dwelling owners become legal owners and secure their assets. Formal and legal ownership also provides an opportunity for upwards that owning a home is an effective means of accumulating wealth among low-income (Herbert et al., 2013), which might be the reason why some people support urban transformation projects. Lastly, urban transformation projects are tools to transfer the lower income group's informally owned properties to a powerful coalition of contractors, public servants, and wealthy consumers (Kuyucu, 2020b). The projects integrate countless people into the formal market by formalizing informal housing settlements (Candan and Kolluoglu, 2008; Demirtas-Milz, 2013; Elicin, 2014; Lovering and Türkmen, 2011). According to Rolnik (2013), two criteria of FoH are privatization of homeownership which includes the sale of public housing, and the extension of housing finance through financing homeownership. Both of these criteria are met in the Turkish case. The social housing program aims to make low- and middle-income groups homeowners who cannot own housing under the current market conditions. TOKİ has title deeds of the social housing; with the sale process, the housing loses the feature of being public property and becomes privatized that the title deeds are not issued until full repayment of the debts. Also, there is no strict control of housing sales in the market. It is still possible to buy and sell a house before construction starts[8] or without a residence permit.[9] Residential development and urban transformation projects keep pushing demand for homeownership because of the demand; payback times are shorter in housing projects than other real-estate or infrastructure projects. Socio-cultural habits are deterministically dominant in housing behavior in addition to factors such as interest rates, income, and other macro-economic conditions. Turkey's housing regime is very similar to the Mediterranean welfare model; families support and help purchase housing. Homeownership is at the center of housing policy—the centrality of owner-occupied tenure cause path-dependent policymaking, and residential property has been perceived as a financial asset. Consideration of housing as the safety box that protects the value of money the most due to the increase in the house price under high inflation conditions also affects motivations of homeownership.

Cultural attitudes towards homeownership also play an essential role in the Turkish case. The rental option has been socially constructed as an insecure type of tenure. In contrast, owner-occupancy is constructed as the best option for the security of tenure. Homeowners are accepted financially and ontologically secure. Homeownership converts technical, social capital, and time into housing for current and future households, successive generations. In market-driven forms of social production of homeownership, it is the family's responsibility. In availability of market channels and the availability of non-market channels to homeownership enabled the expansion of family and family-supported homeownership. Paternalistic thought of providing a home (primarily to the male child), housing arrangements of the couples at or before the time of marriages, supporting children buying a house with under-mattress money or gold (especially for low-income families), intergenerational transmission via the law of decedent's estate are some of the main ways of family support (Arslan Taç, Forthcoming).[10] Rising house prices and the availability of banking finance also affect the level of family support. Even though Halıcıoğlu (2007) finds income is the most critical factor for housing demand in Turkey, the desire to become a homeowner is irrespective of financial situations[11] and still strong because there is a traditional perception of housing: profitable, safe investment in maintaining the purchasing power of money. Although housing and ownership costs are very high for low-income households, it also has a social security role.

On the contrary, to Global North countries like Germany, the United Kingdom, Ireland, there are no mega-landlords in Turkey. Most homeowners are invisible amateur, individual small investors who are multiple homeowners. The context of the rental sector in Turkey differs from other countries. Rental housing is provided mainly by these multiple homeowners rather than institutional landlords. Besides, there is a developed private rental sector in Turkey; due to institutional, historical dynamics, and legacies, the tenancy is fundamentally an urban phenomenon (Balamir, 1999). The existence of built-to-rent housing by institutional investors is not observed in Turkey yet.

Varieties of Residential Capitalism in Turkey: (Inter)National Political Economy of Housing

VoC and RC literature emphasizes how the capitalist system functions differently and focuses on expanding capitalism typologies (see Coates, 2005; Erol, 2019; Fuller et al., n.d.; Hall and Soskice, 2001; Pitcher, 2017; Schwartz and Seabrooke, 2008, 2009). VoC literature is studied in developed economies, GN, mostly (Becker, 2013). In this section, drawing a model for RC in Turkey is not aimed. Rather development of the construction, RE, and housing sector and FoH in Turkey will be illustrated with the insights of RC literature. Turkey tried to increase integration into the world economy through Washington Consensus policies like other developing countries. Turkey can be classified as a state-led market economy that infrastructure projects are transformed into liquid investment objects, like typical emerging economies as Brazil, India, and China. The availability of foreign credit is one of the common features of developing countries. Still, along with Hungary, Indonesia, and Peru, foreign investors hold almost double the total outstanding bonds (McKinsey Global Institute, 2015).

Same Story or Different Trajectory: Financial Architecture of Turkey

The financialization process in the Global South is constrained by structural factors. Although the classification is based on homogeneities, the financialization process in the Global South is more diverse and variegated. There are differences at the national level regarding financial architecture, macro-economic conditions, a coalition of local actors affecting modes of housing provision, and house prices. The roots of the transformation of Turkish capitalism can be traced back to the 1930s when big construction firms were established and became the main instrument of capital accumulation that; in the process of capitalization and urbanization, construction investments constituted up to 70% of fixed capital investments (Tekeli and İlkin, 2004). Political and economic dynamics like Marshall Plans, joining NATO and World Bank credits in the 1950s triggered infrastructure movement, including building new water dams, harbors, and railways, which caused a turning point for the relationship between construction capitalists and macro-economic outcomes and political elites. As an emerging, late-capitalist economy, Turkey has heavily dependent on external capital inflows and credit stimulus. Being overly dependent on capital inflows and manufacturing exports reflected in the shock created by the global financial crisis that Turkey and Mexico hit hard with countries in the (semi) peripheries in Eastern Europe and East and Southeast Asia (Tyson and McKinley, 2014) since the 1990s. One of the indicators of dependent economic integration is foreign capital banks (Table 7.1). The entry of foreign bank branches into the domestic financial industry is part of international financial integration, impacting consumption, investment, growth, and short-term capital flows. The number of domestic banks and foreign banks is almost equal in Turkey. There are 21 foreign deposit banks, 16 of them founded in Turkey, and four foreign development and investment banks[12] (The Banks Association of Turkey, 2020). The ratio of consolidated foreign claims and private credits to the GDP of the banks has been continuously increasing (Table 7.1) in Brazil, China, India, Mexico, and Turkey (The World Bank, 2019b). Turkey's subordinate position, and debt-fueled high growth rates, the share of housing and construction sectors in GDP[13] is incoherent to Global South. Credit-driven economic growth can directly be observed in the housing market.

Table 7.1 Private credit by deposit money banks and other financial institutions to GDP (%)

	1996	2000	2005	2010	2015	2017
Brazil	40.13	30.20	28.53	55.40	70.17	63.78
Mexico	16.83	14.57	14.75	14.75	29.57	29.57
Argentina	18.03	22.44	9.35	11.00	12.54	13.38
Turkey	15.86	13.69	17.66	38.69	61.63	65.11
China	80.38	104.83	106.30	115.67	139.92	150.59
India	21.03	25.64	35.26	44.05	49.79	47.47
United States of America	131.00	162.93	179.53	181.57	178.88	178.88
United Kingdom	96.41	107.56	136.77	186.72	132.47	131.70

Source: Global Financial Development Report 2019/2020 (The World Bank, 2019b) (table prepared by the author).

The share of housing cooperatives almost disappeared from housing provision,[14] and tolerance to squatter policy has abandoned during the 2000s. Housing credits become widely available at the same time. Housing credits' growth rate exceeded 200% annually in the aftermath of the 2001 crisis due to large capital inflows (Tunç, 2020b). Housing and homeownership are constructed as great sources of investment. Even though house prices are rising—the highest annual price increase compared to the quarter of the year, 12-month change is 30.3% (nominal) (Knight, 2021) and 13,5% (real), the profitability of housing investment is in decline. The Turkish economy has been in turbulence since 2018, and the Lira is continuously falling despite upward pressure on prices and interest rates. Therefore, construction costs and housing prices display a constant increase. However, they are dramatically low when converted back to US Dollars. The State actively supports entry into homeownership: housing provision and policies to sustain the accumulation of assets and recycling liquidity to underpin homeownership. Since the 2016 coup attempt, housing sales are very wobbly. Justice and Development Party (AKP) government announced several housing campaigns in cooperation with the sector's biggest actors to dissolve the stock in branded housing projects. With the motto "Become a homeowner just like paying rent", interest rates were reduced, down payments zeroed, and maturities were extended. Pre-cautionary savings are necessary to become a homeowner. However, this is changing due to decreased down payments and an increase in loan terms in mortgages.

New Political Economy of Housing: Where Turkey Stands in the Global Financialization of Housing

Even if there are similarities, there is not one single trajectory of FoH. Private debt data is one of the indicators of housing financialization (Fernandez and Aalbers, 2019). The generalization of credit card usage is part of the ongoing transformation of welfare and the capitalist system of Turkey. Debt and breadth of FoH in Turkey have not reached levels comparable with the cases of GN. There is a growing mortgage finance literature focusing on the positive and negative sides of mortgage finance expansion, mainly related to finance and growth literature and access to housing finance (see Campbell, 2006; Lapavitsas, 2009; Schwartz and Torous, 1989; Tomaskovic-Devey and Lin, 2011; Fernandez and Aalbers 2016; Calza et al., 2013; Badev et al., 2014). Mortgage finance is important because it is a prime example of housing finance. The mortgage debt to GDP ratio of Turkey is behind many countries (Table 7.2). The absence of subprime markets differentiates Turkey from other Global South countries; capital markets replacing banking in housing finance do not exist in Turkey. The market is dominated by domestic, commercial banks (Aslan and Dinçer, 2018; Yeşilbağ, 2020). The central role of the State in financialization is a crucial common trend. In making mortgage lending via public institutions and public funding, Turkey, Mexico, and Taiwan demonstrate the State's fundamental role. Marois (2012) conceptualizes Turkey's finance-led transition similar to Mexico's transition to neoliberalism with emerging market capitalism specific to peripheral capitalism. The direct intervention of the State shows the political importance of housing, especially homeownership in these societies. Financial tendencies are theoretically linked to financialization in GN, foreign capital inflows, banks' shift towards profit-making lending towards households for mortgages, and consumption.

There is a rapid increase in housing loans—the aftermath of the 2001 financial crisis. Housing credit's growth rate exceeded 200% annually—although there is a continuous increase in all loan types (Table 7.3, Figure 7.1). Despite the State's support, mortgaged house sales are 32% of total house sales between 2013 and 2020; household debt-to-income ratios are comparatively low in Turkey. Some contractors provide their payment schemes via bill obligatory or accept a certain amount of cash before the construction starts or accept loans for a certain percentage of the sale price. Based on data for the second quarter of 2020, the most preferred payment is cash in advance (GYODER, 2020b). Changes in housing loan structures are closely related to Turkey's accumulation model and explain how the housing industry/finance is at the center of the Turkish economy and why the State supports housing supply and owner-occupied tenure form. Tunç (2020b) finds that exogenous expansion in housing credits and consumer credits utilized has a large and significant effect on house prices.

Since the 2000s, the Turkish government is trying to attract global investment to enable economic growth and prevent unemployment by manipulating the existing housing regime, normalizing the FoH, privatization, and commercial construction implemented by violating many judicial and master plan decisions.[15] The share of real-estate acquisitions in a total of $8.8 billion foreign direct investment in the first eleven months of 2019 was 52% (GYODER, 2020a). The number of foreign capital companies operating in the construction and real-estate sector is around 10 thousand at the

Table 7.2 The mortgage market in selected countries

		2007	2010	2015	2018
Brazil	Interest rates on new residential loans[a]	n/a	10.30[b]	10.10	8
	Average amount of a mortgage granted[c]	18,210	40,662	38,763	37,061
	Total outstanding residential loans[d]	16,790	59,210	115,872	134,780
	Gross residential loans[e]	9,391	34,901	34,678	27,190
Turkey	Interest rates on new residential loans	18.30	11.05	12.31	19.3
	Average amount of a mortgage granted	25,247	26,643	22,530	19,247[f]
	Total outstanding residential loans	17,500	28,429	45,096	31,023
	Gross residential loans	8,696	15,939	15,464	6,872
United States of America	Interest rates on new residential loans	6.34	4.69	3.83	4.65
	Average amount of a mortgage granted	n/a	n/a	n/a	n/a
	Total outstanding residential loans	8,114,598	7,822,706	9,232,840	9,496,699
	Gross residential loans	1,173,076	1,229,530	1,563,767	1,886,876
United Kingdom	Interest rates on new residential loans	5.69	3.81	2.62	2.11
	Average amount of a mortgage granted	185,571	139,886	199,016	217,692
	Total outstanding residential loans	1,576,978	1,392,970	1,755,387	1,575,990
	Gross residential loans	521,381	155,981	305,547	303,434

Notes:
[a] Annual average based on monthly figures, %.
[b] Figure for 2010 is not available, the number represents 2011.
[c] EUR, the conversion to euros is based on the yearly average bilateral exchange rate.
[d] Total amount, end of the year, EUR million.
[e] Total amount, EUR million.
[f] Figure is not available for 2018, the number represents 2017.

Source: HYPOSTAT 2019 (European Mortgage Federation, 2019) (table prepared by the author).

end of 2019[16] (Ministry of Industry and Technology, 2019). For example, Blackstone[17] invests in any real-estate worldwide, including pension funds and secondary funds, expanded their business to Global South after the housing burst in 2008. In 2008, Blackstone purchased 5,000 public housing, which cost €12 billion, from Madrid's local government (Anti-Eviction

Table 7.3 Total credit to households (core debt) as a percentage of GDP

	2015	2016	2017	2018	2019	2020 (Q1)
Brazil	28.5	28.1	28.4	29.2	30.5	30.4
Mexico	15.2	16	16	16	16.2	16.4
Argentina	6.4	6	6.9	6.7	5.4	5
Turkey	17.9	17.6	17.1	14.7	14.8	15.1
China	38.9	44.2	48.1	51.5	55.2	57.2
India	9.9	10.1	10.8	11.3	12.2	12.6
United States of America	77.6	77.7	77.4	75.7	75.2	75.2
United Kingdom	83.2	83.5	83.2	83.8	83.9	84.5

Source: Credit to the non-financial sector (Bank for International Settlements, 2020) (table prepared by the author).

Mapping Project, n.d.; International Alliance of Inhabitants, 2015). The same year, Blackstone entered the Turkish market and invested in shopping malls around €200 million in 2012 in Ankara, Manisa, and Erzurum, previously owned by Netherlander Redevco (Marirott, 2012). The share of real-estate acquisitions—purchases in revenues—in foreign direct investments increased from 20% to 65% in the 2010–2020 period (GYODER, 2020a).

Urban real estate, especially housing construction, privatization of public land, and mega infrastructure projects such as the new bridge over Dardanelles, Akkuyu Nuclear Power Plant, tunnels, highways, and city

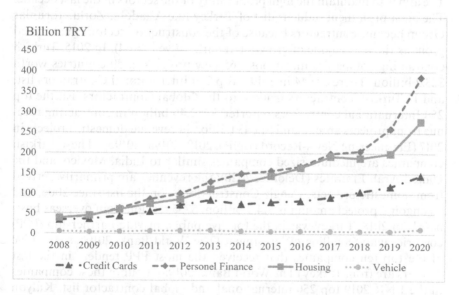

Figure 7.1 Development of consumer loans in Turkey (2008–June 2020).

Source: Prepared by the author based on the main indicators report (Banking Regulation and Supervision Agency, 2021).

hospitals are vital tools of residential capital in Turkey. The total worth of 3rd Airport, Akkuyu Nuclear Power Plant, Eurasia Tunnel, İstanbul Finance Center, North Marmara Highway, and 3rd Bosphorus Bridge, Trans Anatolian Gas Pipeline, Canal İstanbul, İstanbul–İzmir Highway exceed $100 billion. Before 2023, $325 billion is expected to be invested in a set of infrastructure investments as public and PPP projects (Hatem and Akbaş, 2017). These projects are key engines of economic growth, which also serve the reproduction of space. Most of the megaprojects are located in İstanbul to attract global investment. Third Bosphorus Bridge, Eurasia Tunnel, Third Airport, Canal Istanbul have been adopted and in various completion stages. Once construction projects start, they are very likely to continue and splash around. The lands around the Third airport and Canal İstanbul have been purchasing by different capital groups[18] to develop housing and shopping mall projects that megaprojects have created residence for millions. In the last decades, FoH is observed in rural and seaside areas besides seasonal gentrification. The State plays an active role via Presidential Decree on construction and settlement in protected areas or changing national parks' status such as Lake Salda and Uzungöl; thus, develop housing projects in protected areas is possible. Although megaprojects are discussed more due to their scales, housing production constitutes 60% of the domestic construction industry (Şat Sezgin and Aşarkaya, 2017).

AKP continuously supports the construction and real-estate sector with tax cuts, cheaper long-term loans, and favorable regulatory frameworks. The aim is to maintain the high profitability of the sectors in the last decade. The most prominent industrialist of Turkey like Anadolu, Zorlu, Torunlar Group became contractors because of the construction sector's guaranteed profit as the new engine of growth (Akyarlı Güven, 2014). In 2015, Turkish contracting companies undertook 265 new projects in 59 countries worth $23.2 billion. There are 44 in 2019's Top 250 International Contractors list, and 14 Turkish contractors made it to the Global Contractors list; the top 250 International contractors reported $482.40 billion in contracting revenue from projects abroad and $1.043 trillion in revenue domestic projects in 2017 (Engineering News-Record (ENR), 2018, 2019a, 2019b,). These Turkish companies are medium-sized companies similar to India, Mexico, and the United Arab Emirates (Deloitte, 2020); the revenues are primarily derived from domestic projects. Cengiz Construction is on the list since 2006, and its ongoing project amount is $14.2 billion, $735 million from overseas business.[19] In 2018, Turkey was in the top five public-private-partnership (PPP) countries with $8 billion of investment (The World Bank, 2019c: s. 12). Five of the top ten companies that received the most PPP tenders in the last 20 years are from Turkey (The World Bank, 2019a). Four of these companies in the ENR 2019 top 250 international and global contractor list. Kalyon Construction, one of these five companies, was granted a tax exemption of approximately 9.5 billion TL in September 2020 (Presidency of the Republic of Turkey Official Gazette, 2020), the same amount as the High Standard

Railway tender won in August 2020. The Turkish case is characterized by extensive state involvement in the construction sector, as highlighted for the FoH processes in the Global South (Bonizzi, 2013), an example of state-orchestrated financialization. The State drives construction and housing finance (Yeşilbağ, 2020). The macro-political atmosphere in metropolitan cities is highly centralized; local actors cannot effectively implement urban policy planning. Mega housing, infrastructure, construction projects are funded, orchestrated, securitized by the State. The State connects money and power to find new ways, projects to reshape cities, manage these projects with personal connections, and favors powerful market actors by managing land and social housing privatization, large-scale UTPs, mega infrastructure projects, and private toll roads.

Aforementioned, the financialization process in the peripheries mostly depends on capital inflows. The capital does not necessarily flow only from Global North to Global South; the pattern is more grift. Hence, the quantity of inflows and their quality is crucial in *subordinate financialization* that domestic corporations turn to global markets (Powell, 2013). Turkish contractors have been carrying their experiences abroad primarily to developing or less developed countries. From 1972 to 2019, Turkish contractors have undertaken 10.147 projects in 126 countries, with a total value of $401.3 billion (Turkish Contractors Association, 2020). Distribution of international works by country and activities show differences according to the periods. For example, within the period 1980–1989, housing was the number one activity with a share of 40.2%. In 2010–2019 housing is the third biggest (12.1%). Kazakhstan is a recent example of the rapid economic growth grift nature of FoH. The government's decisions on constructing a new capital and changes in housing policy attracted foreign investment (Rolnik, 2019: 268–269). Kazakhstan is the fifth country where Turkish contractors work the most.[20] The projects' total value between 1972 and Q3/2018 is $22.7 billion (Turkish Contractors Association, 2018). The construction sector in Kazakhstan was the first to suffer in the global financial crisis in 2008. Several construction firms were forced to shut down and left the country without completing projects; most firms were from Turkey (Rolnik, 2019, 281). Turkish contractors operate in the international and global markets, develop partnerships, and purchase European contractor companies to expand their capacities.[21] Turkey is experiencing RC and FoH concerning its subordinate position, while Turkish contractors function to spread financialization to the periphery, mainly the Middle Eastern countries and the Russian Federation.

A Brief Review of Policy Tools to Postpone the Housing Crisis in Turkey

Since housing finance is one of the crucial pillars of the economy, a financial crisis in the housing and construction sector can shake the foundations of the economy. The attempts to postpone a potential crisis are ranging from giving extensive authority to TOKİ to integrating housing finance into Sukuk,[22] from selling Turkish Citizenship via *Golden Visa* to promoting housing projects in international real-estate fairs. Another component of FoH and RC, secondary markets, and mortgage securitization are nonexistent in the Turkish case. Although the mortgage market is relatively tiny, Real Estate Investment Trust (REIT) is highly developed in Turkey. REITs are crucial components of FoH and RC since they are designated to fund real-estate projects. REITs were introduced in Turkey in 1995 and experienced remarkable expansion after 2008. There are 33 REIT companies with 36.5 million TL and 75.5 million TL transaction value (GYODER, 2020b) that Turkey ranked. Emlak Konut REIT, privity of Ministry of Environment and Urbanisation, and TOKİ is a market leader since its creation in 2010 (Yeşilbağ, 2020). Sumer and Ozorhon (2020) find that Turkish REITs perform better than global gold prices.

Foreign nationals are buying immovable property in Turkey is facilitated by the amendment made in 2012. In 2019, the real-estate acquisition requirement—one of the exceptional cases in Turkish citizenship acquisition—reduced from $1M to $250,000 or equivalent foreign currency or TRY to revive the stagnant housing market. TOKİ, Emlak Konut REIT, municipalities with other private companies participate in international real-estate exhibitions such as MIPIM, Expo-Real, Cityscape, and the International Property Show in Dubai attract investors in the housing market. These fairs attract impressive crowds of businesspeople and profoundly affect international capital flows and real estate investment. Still, sales to foreigners do not reach the desired level.

Islamic banking began in 1984 in the form of Special Finance Institutions in Turkey and showed no improvement until the 2000s. State-owned participation banks, Turkey Wealth Fund, have been established to develop the Islamic Finance sector after 2002. In addition to these developments, lease certificates, real-estate certificates, and housing bond issuance models were developed to remove interest from the housing finance by making it suitable for Sukuk. Companies are promising "interest-free, saving based finance for homeownership", which is not subject to any legal regulation or supervision, are mushrooming in the last decade. Low-income households who are not eligible to use loans from banks or people who prefer Sukuk are trying to purchase a house and become homeowners with a traditional *gold-day*[23] finance system. These companies offer various payment options, which are interest-free, with or without a down payment, tally payments up to 200 months, a choice to become a group member, and wait to win regular monthly draws or pay tally payments alone. In Argentina and Turkey, informal solutions to overcome the banking system's insufficiency are common (Socoloff, 2020). Although these informal solutions are legally unregulated, they reflect how cultural attitudes can determine housing behaviors.

Concluding Remarks

The 2008 Global Financial Crisis proved that the housing finance system is a significant component of economies at the domestic and global levels. Residential capitalism is path-dependent as FoH and has casual relations with political regimes, political culture, welfare regimes, and financial architecture. Housing as a social institution interacting with financialization is path-dependent. It reflects differences in housing policy, modes of housing provision, pension, social security provision, and financial architecture. This chapter presented the highlights of residential capitalism and FoH of Turkey and argued that Turkey's residential capitalism is based on historical legacies and local dynamics and aimed to show the role of housing with the role of construction in the financial architecture of Turkey. The Turkish case is an example of how FoH and residential capitalism can be uneven, variegated, and heterogeneous. This study is a step to understand VRC in Global South; more research is needed on macro-comparative analysis to capture and to further explore varieties in residential capitalism and FoH.

Acknowledgments

This chapter is based on the fourth chapter author's ongoing PhD dissertation titled "The meaning of home and the cultural motivations of home-ownership in Turkey". I would like to thank to Burcu Özdemir Sarı for her encouragement and thank to the editors and reviewers for their helpful comments.

Notes

1 In periods when housing sales decrease, different campaigns such as 1-year maturity deferral are applied. Commercial banks offer a 240-month fixed rate with a 10% minimum down payment and a down payment recently reduced to 5% to recover COVID-19's effect on the housing market.

2 Famous for its closeness to AKP, Tay Group announced 16:9 in 2011 with $175million investment (Hürriyet Gazetesi, 2011). The administrative court canceled development plans and decided to trim 16:9 in 2013. With the retrial in 2018, the skyscraper became legal (Erbil, 2018).

3 The term entered the political sphere during AKP rule and one of the crucial policy elements of AKP. Each presented to public as dazzling and unprecedented projects that will lead to economic growth and the leading position in the future.

4 These rates are not absolute, reflect estimated data. Construction activities are more concentrated in these metropolitan cities; housing prices are the highest.

5 Unlike the predecessors, AKP named this regulation as *development peace* instead of development amnesty.

6 Istanbul is in the first place, with 1.9 million independent units followed by Izmir with 672,000 and Ankara with 361,000. Type of the buildings is not shared with public.

7 UTPs also lead to gentrification and displacement. The former owners, who cannot afford the price difference caused by the m² difference between demolished and reconstructed houses, have to move to different parts of the city. Similarly, tenants are displaced because they cannot afford rents in these regions anymore.

8 The housing market has been professionalizing since the beginning of the 2000s. However, socio-culturally and politically constructed conducts of informality continue to be practiced. There is a faith and trust that these kinds of informal purchases are safe.

9 Resident permit is not a must for mortgage, banks lend credit if the house has construction permit. Especially in branded housing projects, the construction company signs a housing loan agreement with the bank, and it is possible to use a loan even before the project starts.

10 One of the key findings of the author's dissertation.

11 Due to macro-economic conditions such as high inflation and depreciation of money, housing is not a profitable investment tool as it is believed. Since 2016, TRY has been experiencing a constant loss of currency compared to developing countries. It would be more profitable if the same amount of money was invested in foreign currency on date of house purchase.

12 Although the number of foreign banks in Turkey is relatively high, only two of top ten leading banks by total assets are foreign. There are 6 state-owned banks in total of 48 banks and they have the most assets.

13 The share of the real-estate sector including housing in Turkey's GDP for 2002–2017 is 8.6%, the share of the construction sector for the same period is 6.7% (SBB, n.d.).

14 In 1992 almost 23% of buildings were built by cooperatives, whereas the ratio is 1.2% for 2019 (TURKSTAT, 2020).

15 Çamlıca Mosque and Zorlu Center in İstanbul and passenger guaranteed airports are some examples.

16 This corresponds to approximately 15% of the total foreign capital companies.

17 An American equity fund, one of the largest commercial landlords. The estimated current value of Blackstone's real-estate portfolio by February 2020 is $325 billion (Acker, 2020).

18 Turkish Airlines signed a protocol with Emlak Konut REIT to purchase approximately 1.8 million m² of land near the Third Airport in 2018 (Kamuyu Aydınlatma Platformu, 2018). Sheikha Mozah, the mother of the Emir of Qatar, established a company and bought 44 decares of land on the route of Canal Istanbul (Gazete Duvar, 2019).

19 The ongoing overseas projects are mega infrastructure projects in Duhok-Northern Iraq, Kazakhstan, Croatia, Bulgaria, Slovenia, Kuwait, and Bosnia and Herzegovina (Cengiz İnşaat, n.d.).

20 Twenty-five of 120 internationally operating contractors registered to the Turkish Contractors Association do business in Kazakhstan.

21 Rönesans Holding, a tenth largest international contractor in Europe, acquired Ballast Nedam, a Dutch construction company, in 2015 (Rönesans Holding, n.d.); OYAK Cement acquired all shares of Portuguese Cimpor in 2018 (OYAK, n.d.); Zorlu Energy purchased the shares of Israeli Adnit RE's shares in the ratio of 42.15% in 2019 (Zorlu Enerji, n.d.).

22 An Islamic financial certificate, securities and bonds that comply with Islamic religious law, Islamic finance, interest-free banking principles.

23 Traditional socializing ritual hosted by a group of women and has an economic function. A habit when having a bank account was relatively odd that women give cash or gold to the host to get equal gold or money when it's their turn. The meetings can happen at different frequencies.

References

Aalbers MB, Rolnik R and Krijnen M (2020) The financialization of housing in capitalism's peripheries. *Housing Policy Debate* 30(4): 481–485.

Acker D (2020) How Blackstone became the world's biggest corporate landlord. *Fortune*. Available at: https://fortune.com/2020/02/17/blackstone-commercial-real-estate-business-brep-breit/. Accessed 8 March 2020.

Aksu Çam Ç (2020) *Türkiye'de konut siyaseti*. Ankara: Siyasal Kitabevi.

Akyarlı Güven A (2014) Türk Sanayicileri Anlattı: Neden Müteahhit oluyoruz? *Wall Street Journal*. Available at: https://www.wsj.com/articles/turk-sanayici-anlatt-neden-muteahhit-oluyoruz-1407324912. Accessed 24 August 2020.

Allen J, Barlow J, Leal J et al. (2004) *Housing and Welfare in Southern Europe*. Oxford: Blackwell.

Anti-Eviction Mapping Project (n.d.) Properties Owned by Blackstone in CA. Available at: http://www.antievictionmappingproject.net/blackstone.html. Accessed 8 March 2020.

Arslan Taç Z (Forthcoming). *Türkiye'de konut edinimi, ev sahipliğinin kültürel motivasyonları ve evin anlamı*. PhD dissertation, Marmara University, Istanbul, Turkey.

Arslanalp M (2018) Coalitional politics of housing policy in AKP's Turkey. *POMEPS* 31: 25–33.

Ashman S, Fine B and Newman S (2011) The crisis in South Africa: Neoliberalism, financialization and uneven and combined development. *Socialist Register* 47: 174–195.

Aslan AS and Dinçer İ (2018) The impact of mortgage loans on the financialization process in Turkey. *Planlama* 28(2): 143–153.

Badev A, Beck T, Vado L and Walley S (2014) Housing Finance Across Countries. Policy Research Working Paper No. 6756. Available at: https://openknowledge. worldbank.org/handle/10986/16821?locale-attribute=en. Accessed 5 November 2020.

Balamir M (1975) Kat mülkiyeti ve kentleşmemiz. *ODTÜ Mimarlık Fakültesi Dergisi* 1(2): 295–318.

Balamir M (1999) Formation of private rental stock in Turkey. *Journal of Housing and the Built Environment* 14(4): 385–402.

Bank for International Settlements (2020) *Total credit to households (core debt)*. Available at: https://stats.bis.org/statx/srs/table/f3.1. Accessed 1 October 2020.

Banking Regulation and Supervision Agency (2021) *Main indicators report*. Available at: https://www.bddk.org.tr/Veri/Detay/171. Accessed 1 March 2021.

Başlevent C and Dayioğlu M (2005) The effect of squatter housing on income distribution in urban Turkey. *Urban Studies* 42(1): 31–45.

Becker U (2013) Measuring change of capitalist varieties: Reflections on method, illustrations from the BRICs. *New Political Economy* 18(4): 503–532.

Bengs C (2015) The housing regime of Sweden: Concurrent challenges—Part A: Aims, effects and interpretations. Aalto University publication series. Available at: http://urn.fi/URN:ISBN:978-952-60-6474-1. Accessed 20 March 2020.

Bonizzi B (2013) Financialization in developing and emerging countries a survey. *International Journal of Political Economy* 42(4): 83–107.

Bonizzi B, Kaltenbrunner A and Powell J (2020) Subordinate Financialization in Emerging Capitalist Economies. In: Mader P, Meretns D and Van Der Zwan N (eds) *The Routledge International Handbook of Financialization*. New York: Routledge, 177–198.

Calza A, Monacelli T and Stracca L (2013) Housing finance and monetary policy. *Journal of the European Economic Association* 11(S1): 101–122.

Campbell JY (2006) Household finance. *The Journal of Finance* 61(4): 1553–1604.

Candan AB and Kolluoğlu B (2008) Emerging spaces of neoliberalism: A gated town and a public housing project in Istanbul. *New Perspectives on Turkey* 39(39): 5–46.

Cengiz İ (n.d.) Devam Eden Projeler. Available at: https://www.cengiz-insaat.com.tr/#. Accessed 9 September 2020.

Coates D (2005) *Varieties of Capitalism, Varieties of Approaches*. New York: Palgrave.

Danış D and Kuyucu T (2020) Benzer Süreçler, Farklı Sonuçlar: Kayseri, Denizli ve Malatya da Kentsel Dönüşüm Projelerinin Karşılaştırmalı Analizi. In: Geniş Ş (ed) *Otoriter Neoliberalizmin Gölgesinde*. Ankara: Nika Yayınevi, 155–198.

Deloitte (2020) GPoC 2019-Global Powers of Construction. Available at: https://www2.deloitte.com/ content/dam/Deloitte/at/Documents/presse/Deloitte-Global-Powers-of-Construction-2019.pdf. Accessed 20 March 2021.

Demirtaş-Milz N (2013) The regime of informality in neoliberal times in Turkey: The case of the Kadifekale urban transformation project. *International Journal of Urban and Regional Research* 37(2): 689–714.

Elicin Y (2014) Neoliberal transformation of the Turkish city through the Urban Transformation Act. *Habitat International* 41: 150–155.

Engineering News-Record (ENR) (2018) ENR 2018 Top 250 Global Contractors. Available at: https://www.enr.com/toplists/2018-Top-250-Global-Contractors-1. Accessed 5 March 2020.

Engineering News-Record (ENR) (2019a) ENR s 2019 Top 250 Global Contractors. Available at: https://www.enr.com/toplists/2019-Top-250-Global-Contractors-1. Accessed 5 March 2020.

Engineering News-Record (ENR). (2019b) ENR s 2019 Top 250 International Contractors. Available at: https://www.enr.com/toplists/2019-Top-250-International-Contractors-1. Accessed 5 March 2020.

Erbil Ö (2018) 16/9 tıraştan kurtuldu. Silueti bozan kuleler yasallaştı. *Hürriyet*. Available at: https://www.hurriyet.com.tr/gundem/16-9-tirastan-kurtuldu-silueti-bozan-kuleler-yasallasti-40714741. Accessed 1 March 2020.

Ergüven E (2020) The political economy of housing financialization in Turkey. *Housing Policy Debate* 30(4): 559–584.

Erol I (2019) New geographies of residential capitalism: Financialization of the Turkish housing market since the early 2000s. *International Journal of Urban and Regional Research* 43(4): 724–740.

European Mortgage Federation (2019) HYPOSTAT 2019 | A Review of Europe's Mortgage and Housing Markets. Available at: https://hypo.org/app/uploads/sites/3/2019/09/HYPOSTAT-2019_web.pdf. Accessed 10 December 2019.

Fernandez R and Aalbers MB (2016) Financialization and housing: Between globalization and Varieties of Capitalism. *Competition and Change* 20(2): 71–88.

Fernandez R and Aalbers MB (2019) Housing financialization in the Global South: In search of a comparative framework. *Housing Policy Debate* 30(4): 680–701.

Fuller G, Johnston A and Regan A (n.d.) *Bringing the Household Back in: Comparative Capitalism and the Politics of Housing Markets*. UCD Geary Institute for Public Policy Discussion Paper Series No. WP201807. Dublin: Geary Institute, University College Dublin.

Gazete D (2019) *Qatar Emir's mother bought land from Kanal Istanbul area, Erdoğan confirms.* Available at: https://www.duvarenglish.com/politics/2019/12/20/qatar-emirs-mother-bought-land-fromkanal-istanbul-area-erdogan-confirms/. Accessed 19 May 2020.

Gotham KF (2009) Creating liquidity out of spatial fixity: The secondary circuit of capital and the subprime mortgage crisis. *International Journal of Urban and Regional Research* 33: 355–371.

Gündoğmuş YN (2019) *İmar barışından 7,5 milyon bağımsız birim faydalandı.* Available at: https://www.aa.com.tr/tr/ekonomi/imar-barisindan-7-5-milyon-bagimsiz-birim-faydalandi/1679907. Accessed 24 June 2020.

Güngen AR (2017) Finansal Taban Yayılma Siyaseti ve Türkiye de Devletin Finansallaşması. In: Bedirhanoğlu P, Çelik Ö and Mıhçı H (eds) *Finansallaşma Kıskacında Türkiye de Devlet, Sermaye Birikimi ve Emek.* İstanbul: NotaBene Yayınları, 23–44.

GYODER (2020a) *Realestate Turkey: A Close Look to Comparable Markets.* Available at: https://www.gyoder.org.tr/uploads/yayinlar/gyoderkasimv3.pdf. Accessed 4 May 2021.

GYODER (2020b) *Türkiye Gayrimenkul Sektörü 2020—2. Çeyrek Raporu—NO:2.* İstanbul. Available at: https://www.gyoder.org.tr/uploads/yayinlar/GOSTERGE-CEYREK2-2020-web.pdf. Accessed 28 April 2021.

Hall PA and Soskice D (eds) (2001) *Varieties of Capitalism: The Institutional Foundations of Comparative Advantage.* Oxford and New York: Oxford University Press.

Halıcıoğlu F (2007) The Financial Development and Economic Growth Nexus for Turkey. *EERI Research Paper Series*, No. 06/2007. Available at: https://mpra.ub.uni-muenchen.de/3566/1/MPRA_paper_. Accessed 21 June 2020.

Hatem E and Akbaş M (2017) Capital Projects and Infrastructure Spending in Turkey Outlook to 2023. Available at: https://www.pwc.com.tr/tr/advisory/capital-project-and-infrastructure-spendingin-turkey-pwc.pdf. Accessed 28 May 2020.

Herbert C, McCue D and Sanchez-Moyano R (2013) Is homeownership still an effective means of building wealth for low-income and minority households? (Was it ever?), *Harvard University Joint Center for Housing Studies.* Available at: https://www.jchs.harvard.edu/sites/default/files/hbtl-06.pdf. Accessed on 20 April 2020.

Hürriyet Gazetesi (2011) *Zeytinburnu'na yeni otel planlıyor.* Available at: https://www.hurriyet.com.tr/ekonomi/zeytinburnu-na-yeni-otel-planliyor-17115239. Accessed 24 June 2020.

Kamuyu Aydınlatma Platformu (2018) *Emlak Konut GYO A.Ş. Özel Durum Açıklaması.* Available at: https://www.kap.org.tr/tr/Bildirim/685864. Accessed 19 May 2020.

Kayasü S and Uzun N (2009) Kentsel Dönüşüm/Yenileme—Kentsel Yeniden Canlandırma/Yenileş(tir)me: Kavramlara Yeni Bir Bakış. In: Kayasü S, Işık O, Uzun N and Kamacı E (eds) *Gecekondu, Dönüşüm, Kent.* Ankara: Orta Doğu Teknik Üniversitesi Mimarlık Fakültesi, 151–161.

Keyder Ç (2005) Globalization and social exclusion in Istanbul. *International Journal of Urban and Regional Research* 29(1): 124–134.

Kızıldağ Özdemirli Y (2018) *Ankara* In: Rocco R and Van Ballegooijen J (eds) *The Routledge Handbook on Informal Urbanization.* New York: Routledge, 22–33.

Knight F (2021) *Global house price index research Q4 2020.* Available at: https://content.knightfrank.com/research/84/documents/en/global-house-price-index-q4-20207884.pdf. Accessed 17 March 2021.

Kuyucu T (2020a) The great failure: The roles of institutional conflict and social movements in the failure of regeneration initiatives, Istanbul. *Urban Affairs Review* 1–35. DOI: 10.1177/1078087420957736.

Kuyucu T (2020b) Türkiye'de Sosyal Konut Politikasının Paradoksu. In: Geniş Ş (ed) *Otoriter Neoliberalizmin Gölgesinde*. Ankara: Nika Yayınevi, 123–154.

Lapavitsas C (2009) Financialised capitalism: Crisis and financial expropriation. *Historical Materialism* 17: 114–148.

Lovering J and Türkmen H (2011) Bulldozer neo-liberalism in İstanbul: The state-led construction of property markets, and the displacement of the urban poor. *International Planning Studies* 16(1): 73–96.

Marirott R (2012) Blackstone Appoints Turkish Business Luminary. Available at: https://www.pere news.com/blackstone-tees-up-first-deal-in-turkey/. Accessed 5 March 2020.

Marois T (2012) *States, Banks and Crisis: Emerging Finance Capitalism in Mexico and Turkey*. Cheltenham: Edward Elgar Pub.

Mawdsley E (2016) Development geography II: Financialization. *Progress in Human Geography* 42(2): 1–11.

McKinsey Global Institute (2015) *Debt and (not much) Deleveraging*. Available at: https://www.mckinsey.com/featured-insights/employment-and-growth/debt-and-not-much-deleveraging#. Accessed 17 May 2020.

OYAK (n.d.) *OYAK Acquired the Portuguese Cement Giant Cimpor.* Available at: https://www.oyak. com.tr/news/oyak-acquired-the-portuguese-cement-giant-cimpor/. Accessed 20 September 2020.

Özdemir Sarı ÖB (2019) Redefining the housing challenges in Turkey: An urban planning perspective. In: Özdemir Sarı ÖB, Özdemir SS and Uzun N (eds) *Urban and Regional Planning in Turkey*. Switzerland: Springer, 167–184.

International Alliance of Inhabitants (2015) *PAH Global action: Blackstone.* Available at: https://www.habitants.org/news/inhabitants_of_americas/pah_global_action_ blackstone. Accessed 6 March 2020.

Ministry of Environment and Urbanisation (2018) *İmar Barışı (Development Peace)*. Available at: https://webdosya.csb.gov.tr/db/imarbarisi/icerikler/brosur-20180603111057.pdf. Accessed 11 August 2020.

Ministry of Industry and Technology (2019) *Yabancı Sermayeli Firma Listesi*. Available at: https://sanayi.gov.tr/merkez-birimi/14a09761d390/diger. Accessed 30 August 2020.

Pitcher MA (2017) Varieties of residential capitalism in Africa: Urban housing provision in Luanda and Nairobi. *African Affairs* 116(464): 365–390.

Powell J (2013) *Subordinate Financialisation: A Study of Mexico and Its Non-Financial Corporations*. PhD Thesis. SOAS University of London, London.

Presidency of the Republic of Turkey Official Gazette (2020) *09 Ekim 2020 Tarihli ve 31269 Sayılı Resmi Gazete*. Available at: https://www.resmigazete.gov.tr/eskiler/2020/10/20201009-1.pdf. Accessed 9 October 2020.

Rolnik R (2013) Late neoliberalism: The financialization of homeownership and housing rights. *International Journal of Urban and Regional Research* 37(3): 1058–1066.

Rolnik R (2019) *Urban Warfare: Housing Under the Empire of Finance*. London: Verso Books.

Rönesans Holding (n.d.) *JV between Rönesans Holding and TAV Construction in Europe*. Available at: https://ronesans.com/en/ronesans-holding-ve-tav-insaat-ortakligi-avrupada-2/. Accessed 20 September 2020.

SBB (T.C. Cumhurbaşkanlığı Strateji ve Bütçe Başkanlığı) (n.d.). Ekonomik ve Sosyal Göstergeler (Economic and Social Indicators). Available at: https://www.sbb.gov.tr/ekonomik-ve-sosyal-gostergeler/#1540021349004-1497d2c6-7edf. Accessed 8 September 2020.

Schwartz H and Seabrooke L (2008) Varieties of residential capitalism in the international political economy: Old welfare states and the new politics of housing. *Comparative European Politics* 6(3): 237–261.

Schwartz H and Seabrooke L (eds) (2009) *The Politics of Housing Booms and Busts.* Basingstoke and Hampshire: Palgrave Macmillan.

Schwartz ES and Torous WN (1989) Prepayment and the valuation of mortgage-backed securities. *The Journal of Finance* 44(2): 375–392.

Socoloff I (2020) Subordinate financialization and housing finance: The case of indexed mortgage loans coalition in Argentina. *Housing Policy Debate* 30(4): 585–605.

Sumer L and Ozorhon B (2020) Investing in gold or REIT index in Turkey: Evidence from global financial crisis, 2018 Turkish currency crisis and COVID-19 crisis. *Journal of European Real Estate Research* 14(1): 84–99.

Şat Sezgin AG and Aşarkaya A (2017) *İnşaat Sektörü.* İstanbul: Türkiye İş Bankası.

Şensoy Boztepe A (2020) *Çevre ve Şehircilik Bakanı Kurum: Yeni imar barışı süreci planlanmıyor.* Available at: https://www.aa.com.tr/tr/turkiye/cevre-ve-sehircilik-bakani-kurum-yeni-imar-barisi-sureci-planlanmiyor/1746423. Accessed 24 June 2020.

Tekeli İ (2009) *Konut sorununu konut sunum biçimleriyle düşünmek.* İstanbul: Tarih Vakfı Yurt Yayınları.

Tekeli İ and İlkin S (2004) *Cumhuriyetin Harcı (3): Modernitenin Altyapısı Oluşurken.* İstanbul: İstanbul Bilgi Üniversitesi Yayınları.

The Banks Association of Turkey (2020) *Grup Bazında, Banka ve Bankaların Şube Sayısı.* Available at: https://www.tbb.org.tr/modules/banka-bilgileri/banka_sube_bilgileri.asp. Accessed 1 October 2020.

The World Bank (2019a) *Featured Rankings, 1990 to 2019.* Available at: https://ppi.worldbank.org/en/snapshots/rankings. Accessed 8 September 2020.

The World Bank. (2019b) *Global Financial Development Report 2019/2020: Bank Regulation and Supervision a Decade after the Global Financial Crisis.* Available at: https://www.worldbank.org/en/publication/gfdr/data/global-financial-development-database. Accessed 22 August 2020.

The World Bank (2019c) *Private Participation in Infrastructure 2019.* Available at: https://ppi.worldbank. org/en/ppi. Accessed 22 August 2020.

TMMOB (2004) Kaçak Yapılaşma İle İlgili Süreçler, Sorunlar, Çözüm Önerileri Değerlendirme Raporu. *Planlama* 3: 95–105.

TOKİ (2018) *TOKİ Nevşehir de "Kentsel Dönüşüm" kapsamında modern bir mahalle inşa ediyor.* Available at: https://toki.gov.tr/haber/toki-nevsehirde-kentsel-donusum-kapsaminda-modern-bir-mahalle-insa-ediyor. Accessed 21 May 2020.

Tomaskovic-Devey D and Lin KL (2011) Income dynamics, economic rents, and the financialization of the U.S. Economy. *American Sociological Review* 76(4): 538–559.

Tunç C (2020a) *Konut piyasası, finansmanı ve göstergeleri.* İstanbul: Litera Yayıncılık.

Tunç C (2020b) The effect of credit supply on house prices: Evidence from Turkey. *Housing Policy Debate* 30(2): 228–242.

Turkish Contractors Association (2018) *Turkish Contracting in the International Market.* Available at: https://tmb.org.tr/doc/file/TCIM_April_2018.pdf. Accessed 12 May 2020.

Turkish Contractors Association (2020) *Turkish Contracting in the International Market*. Available at: https://www.tmb.org.tr/doc/file/YDMH_2019_10april2020. pdf. Accessed 20 May 2020.

Turkish Court of Accounts (2008) Büyükşehir Belediyelerinde Altyapı Faaliyetlerinin Koordinasyonu. Available at: https://www.sayistay.gov.tr/tr/Upload/62643830/files/raporlar/diger/BüyükşehirBelediye lerindeAltyapıFaaliyetlerininKoordinasyonu-PerformansDenetimiRaporu.pdf. Accessed 26 February 2020.

TURKSTAT (2020) *Yapı İzin İstatistikleri*. Available at: https://biruni.tuik.gov.tr/yapiizin/giris.zul. Accessed 17 September 2020.

Türel A and Koç H (2015) Housing production under less-regulated market conditions in Turkey. *Journal of Housing and the Built Environment* 30(1): 53–68.

Tyson J and McKinley T (2014) Financialization and the developing world: Mapping the issues. FESSUD Working Paper Series No. 38. Available at: http://citeseerx.ist.psu.edu/viewdoc/download;jsessionid=FF57B7C12E7B64BEDFA1254A088D6AF8?-doi=10.1.1.676.4181&rep=rep1&type=pdf. Accessed 26 March 2020.

Yeşilbağ M (2020) The state-orchestrated financialization of housing in Turkey. *Housing Policy Debate* 30(4): 533–558.

Zorlu Enerji (n.d.) *Adnit Real Estate Ltd*. Available at: https://www.zorluenerji.com.tr/en/corpo/sunsidiaries-and-asso/adnit-real-estate-ltd. Accessed 20 September 2020.

8 Everyday Negotiations of Finance and Debts in State-Led Mass Housing Estates for Low-Income Households in Istanbul

Esra Alkim Karaagac

Introduction

This chapter aims to provide an ethnographic window into the Turkish State's Housing Development Administration's (TOKİ) housing provision for low-income groups, with debt serving as the analytical cursor across multiple scales: within the households, in between the households, and the state institutions. Consequently, it aims to discuss how *debt to state* is manifested and experienced in the context of state-led housing provision for low-income groups in Turkey. The chapter shares the findings of ethnographic research conducted in Istanbul in a low-income TOKİ mass housing estate[1] during fall 2019 and examines lived experiences of indebtedness to better understand how policy shapes relations of everyday finance, homeownership, and politics. It traces the experiences of households navigating through a private housing market that emerges from state-led housing projects as they mitigate current debt burdens with the hope of future financial gains in real estate.

The purpose is to understand how debt plays a role in establishing consent for the neoliberal transformation of urban land, as well as how housing policy plays a role in establishing consent for debt-based housing provision. The next section provides a brief background of the state-led housing provision for low-income groups in Turkey and sets out the theoretical works on the politics of debt and lived experiences of indebtedness across social disciplines. The following section outlines the methodological approach and the methods used in this research, summarizing how the fieldwork process has unfolded. Section four discusses the negotiations of finance and debt in state-led housing provision in three interconnected themes: provision, debt, and profit. The final section concludes with conceptual and theoretical contributions and future research directions. Overall, the chapter aims to shed light on the indebted lives created in TOKİ mass housing projects, tracing everyday negotiations of debt and theorizing how debts to the state transform the conduct of citizens.

DOI: 10.4324/9781003173670-10

Frameworks

State-Led, Debt-Based Housing Provision
for Low-Income Groups in Turkey

The 2001 debt crisis in Turkey was followed by an IMF imposed recovery program. With a neoliberal agenda, this program boosted the construction sector, facilitating government intervention to transform urban land through housing mobilization programs (Erol, 2019; Lelandais, 2015). In 2003, TOKİ was tasked with developing an estimated six million new housing units across major cities and running a financial model to provide credits for homebuyers with 15–20 years' long payment plans (TOKİ, 2011). Pivotal actors of this housing provision model have been TOKİ, state-run public banks, local municipalities, contractors, real estate markets, and homebuyers. TOKİ produces mass housing projects for the poor, low-income, and middle-income groups, accounting for 86% of total housing provided by the administration, with the remaining 14% being revenue sharing projects with private partners (TOKİ, 2020a). The projects for the poor groups are distributed by the Ministry of Family, Labour, and Social Services, while TOKİ only constructs the houses in those projects. Whereas low-income and middle-income group projects are sold by TOKİ with varying purchase rates and repayment plans. On its official website, TOKİ calls poor, low-income, and middle-income group housing: "social (affordable) housing" (TOKİ, n.d.a). These housing programs target a population segment that the commercial banks do not serve (Pınar and Demir, 2016), offering them the chance to become homeowners by selling flats in remote mass housing estates.

TOKİ's housing provision model is designed around homeownership ideology, excluding other forms of social housing such as free provision, securitizing tenure, or rent subsidizing (Whitehead, 2017). Although the administration claims to "determine the sale prices without a profit purpose, in view of the saving patterns and monthly affordability of the target groups" (TOKİ, n.d.a), the model works on the premise of urban transformation for rent generation, boosting the construction sector, and promoting homeownership. The clients of TOKİ's social housing units either come from inner-city urban transformation projects (UTP) where urban squatter housing residents, evicted and dispossessed, are coerced to resettle in a peripheral TOKİ estate (Türkün, 2011), or through mass housing campaigns, giving homeownership chance to low-income households who struggle to find affordable housing in the city due to ever-increasing rents and housing prices. Either coming through UTPs or mass housing campaigns, access to housing is conditioned upon debt payments.

TOKİ works with public banks (Ziraat Bank and Halkbank) to sign contracts with homebuyers and provide and collect loans on their behalf. TOKİ's housing credits are different from mortgages in the private market, as households "pay directly to the TOKİ account in the public banks with a different price index system" (Doğru, 2016: 210). Furthermore, mortgage insurance and credit, which are important components of the private sector housing finance, do not exist in the TOKİ model (Sarıoğlu-Erdoğdu, 2014). Also, in low-income housing groups, the ownership and title deeds are retained by TOKİ until the debtors pay off their debts (TOKİ, n.d.a). Moreover, the purchase agreement states that if the clients fail to pay their monthly instalments two times in a row, they will face the risk of confiscation and repossession (TOKİ, n.d.b). Over the past 17 years, TOKİ has sold more than 845,000 housing units (TOKİ, 2019a: 10), indebting thousands of households and redefining state-citizen relationship in financial terms; in other words, to long-term "creditor-debtor relationship" (Lazzarato, 2012: 30).

The critical housing literature so far has theorized how neoliberal restructuring of welfare states has shifted the scope of housing policy from urban rights to financial commitments (Aalbers, 2016; Rolnik, 2013; Ronald and Elsinga, 2012). Accordingly, there is an increasing need for qualitative research in TOKİ housing estates to understand how the policy plays in the field, shaping urban fringes as debt geographies and contributing to the growing economic and spatial inequalities. In the Turkish context, studies show that state-led urban transformation and debt-based housing provision, by putting the costs of housing mobilization on urban poor (Lovering and Türkmen, 2011; Türkün, 2015), transform TOKİ low-income mass housing areas into spaces of concentrated poverty and deprivation (Candan and Kolluoğlu, 2008; Kuyucu, 2014). TOKİ's housing regime has been investigated in many respects, such as the privatization of public land and state-led property transfers (Kuyucu and Ünsal, 2010; Türkün, 2011), neoliberalization of housing policies and the oppressive hand of TOKİ in urban transformation (Sarıoğlu-Erdoğdu, 2014; Ünsal, 2015) and unaffordability of the housing projects for the low-income groups (Aslan and Güzey, 2015; Özdemir Sarı and Aksoy Khurami, 2018). Research in these areas offers important insights into the political economy of state-led housing provisions; yet, with notable exceptions (Erman, 2016; Hatiboğlu Eren, 2017; Türkün, 2014), existing research rarely focus on the intimate, situated, and relational aspects of housing debts and the lived experiences of indebtedness to state, in TOKİ housing programs. Accordingly, the power relation between the creditor (as the state) and the debtor (as the citizen) with respect to different functions of debt is underexplored.

Politics of Debt and Lived Experiences of Indebtedness

In the past three decades, structural transformations in government policies from welfare towards the market have led to increasing household debt (e.g., mortgage, medical, student debts) all over the world. An emerging body of literature examines this global debt accumulation and draws attention to its relationship with the authoritarian forms of control (Di Muzio and Robbins, 2016; Graeber, 2011; Lazzarato, 2015). As the discourse of social rights gives way to financial commitments, it empowers governments to act in anti-democratic ways (Tansel, 2017), to secure the financial power of lending institutions at the expense of the public (Lazzarato, 2012), and to evade opposition and resistance, especially for low-income groups who fall into debt earlier in life and never escape the circle of debt (Martin, 2015; Peters, 2016). Debt obligations have a profound disciplining effect on individuals, creating financial subjectivities that can be governed by financial means (Di Feliciantonio, 2016; García-Lamarca and Kaika, 2016; Kear, 2013). The debt economy today is depriving the vast majority of North Americans, Europeans, and people in developing countries of political power and freedoms through rising household indebtedness and financial insecurity (Di Muzio and Robbins, 2016; Montgomerie, 2013; Soederberg, 2014).

As an example of this global trend, since the early 2000s in Turkey, the government's agenda has shifted the scope of public policy from rights to debt (Ergüder, 2015; Özdemir, 2011; Türkün, 2015). Between 2002 and 2020, the total debt per household has gone up 88 times, while the minimum wage increased just twelvefold[2] (AÇSH, 2021; CBRT, 2006: 12–17, 2020a: 25; TEK, 2003: 6; TURKSTAT, 2020). The debt to banks accounted for the lion's share of overall household debt (92.6% in 2020), which included mortgages, auto loans, consumer credits, and credit card debts (CBRT, 2020a). Over the same period, Turkish household debt to GDP tripled (from 6% to 18%), while household debt to disposable income ratio increased more than elevenfold (from 4.7% to 53%) (CBRT, 2006: 12–17, 2020b: 1–13). That means, with increasing dependency of basic consumption on credit, households have borrowed more and more to get by and maintain a living. During the pandemic, gross household debts hit an all-time high of 883.2 billion Turkish Liras (TRY), with 34 million people indebted to the financial system (TBB, 2020). Moreover, the "extension of consumer credit, especially to low-income households and wage earners" (Karaçimen, 2014: 162) has led to financial burdens and debt stress to become prevalent in the everyday lives of low-income households.

While debt is often conceptualized in a linear fashion—as repayment by borrowers whose character and identity remain fixed—social researchers recognize the porosity and openness of social life to debt that is shaped by the ebb and flow of debt-relations (Hall, 2016; Harker, 2017). Debts are embodied processes that are cared for within and through households, not just financial commitments that are strictly handled in terms of incomings and outgoings (Karaagac, 2020; Montgomerie and Tepe-Belfrage, 2017). In reality, debt is not only an abstract, numerical entity that one owes but also something that one lives through as sustaining life (Han, 2012; Joseph, 2014). Therefore, in order to mobilize against systematic indebtedness caused by neoliberal regulations, one needs to challenge dehumanizing and depoliticizing narratives over global debt regimes by tracing everyday debt negotiations as shared experiences (Federici, 2014). Negotiation entails unequal power dynamics and results in coercion, as often as consent. Nonetheless, power, although not equally, is "dispersed through the society" (Few, 2002: 31) and often operates in both ways: dominance and resistance (Sharp et al., 2000). That means citizens who are often seen as passive recipients of certain authoritarian policies also have a certain amount of agency to negotiate their situation and survive. Similarly, TOKİ housing provision results not only in financial responsibilities that must be met but also in political and spatial conflicts that are negotiated at the home, housing site, and institutional scales. The next section expands on the research design with respect to these conflictual spaces and scales of TOKİ housing provision and debt negotiations.

An Ethnographic Lens to TOKİ Housing Estates: How Does Policy Play in the Field?

The research draws on the findings of four months long fieldwork in Turkey between August and December 2019, using multiple qualitative methods. It opens an ethnographic window to state-led housing provision for low-income groups. While the research is centred around a low-income mass housing estate in Istanbul, it also involves interviews with TOKİ officials in its headquarters and experts in public banks. The housing site where this research has been conducted was built in 2009, in three consecutive phases. Today it is located at the border of Istanbul with the neighbouring city Kocaeli, approximately 36 km and two hours away from the city centre of the Anatolian side (Kadıköy) by public transport. There are about 3,100 housing units in this mass housing area, built in the form of 71 high-rise blocks. The first two phases of the project were built for the low-income (1,744 units) and the last phase for the middle-income groups. The housing units in all three phases were allocated by receiving applications and drawing lots in 2010. This ethnographic research focusses on the low-income housing estates, predominantly populated by households who migrated to Istanbul from the rural areas of Black Sea provinces in northern Anatolia. The majority arrived in Istanbul at a young age and lived in various parts of the city before relocating to TOKİ. To protect the study participants being targeted by authorities, the name of the housing estate is removed from the chapter and pseudonyms are used to help keep individuals' identities confidential.

During the fieldwork, one-on-one and in-depth interviews were conducted with 59 TOKİ residents. Interviews examined individuals' and households' everyday financial practices, strategies to negotiate debts with authorities, and financial aspirations regarding their houses in an emerging housing market. Also, these interviews gathered personal data on how housing debt to TOKİ impacts household members' employment relations, household responsibilities, and future investment plans. Multiple individuals in certain households have been interviewed to understand the gendered and generational aspects of indebtedness to the state. A purposeful sampling method was used in recruitment to select the most appropriate households based on the research questions. Concurrently, one-month-long participant observation was conducted at the housing estate management office,[3] playgrounds, women's weekly gatherings in the building shelter units for the Friday prayers, and men's casual gatherings in the local coffee houses.

Multiple qualitative methods provided insight into different ways that households discuss the policy processes that make them indebted to the state for a house and how they see these processes shaping their relationships to labour and housing markets. In addition, semi-structured interviews with local authorities (local chief,[4] planners in the local municipality, public bank officials, and the housing estate management) helped gather information on their role as an intermediary between the state and TOKİ households in facilitating policy implementation and resolving financial disputes. Moreover, interviews with local real estate agents aided in comprehending the housing market dynamics and their narratives of the future of this TOKİ site in relation to current real estate trends and household behaviours. Additionally, interviews with TOKİ officials at the headquarters aimed to help understand their narrative and rationale for debt-based housing provision for low-income groups. This research has been designed and carried out by the author, who has singly conducted all the interviews and participant observation at the TOKİ housing site. The following section draws on the content and textual analysis of the transcribed interview texts, fieldwork memos, participant observation notes, policy documents, as well as the statistical analysis of numerical data collected.

Everyday Negotiations of Finance and Debt

This section examines TOKİ housing provision and TOKİ housing areas as the hubs of finance and debt negotiations. Negotiation in everyday life unfolds in micro details and subtle ways. Accordingly, the section theorizes everyday negotiations in three respective areas: (1) housing provision and race for homeownership, (2) debt management and financial responsibilization, and (3) profit-seeking in a quasi-real-estate market.

The Road to Homeownership: Worthy Enough in the Eye of the State

The inadequacies in the urban affordable housing market and inaccessibility of housing finance for low-income groups put them in a position to lean back on "the 'benevolent hand' of the Turkish state" for a decent house (Doğru, 2016: 1). The majority of TOKİ households interviewed mentioned they had no choice but to apply for TOKİ campaigns in order to own a house, particularly given rising housing prices and stagnant wages. One of the participants, Asude says

"We struggled for 3-4 years to buy a house before TOKİ. We didn't have the (financial) power. We came up short on money in each attempt. Then I heard about TOKİ projects on TV. I begged my husband to apply for it. He was hesitant about it, he has always been scared of debt. But, what else could we do, creep in rentals all our lives?"

(52, babysitter)

Even the application process, however, becomes a physical and financial strain for some, as applicants must wait in line for days and pay an application fee as much as two months' instalments of a new housing unit (in the year 2008). Interviewees said that they had to borrow from their relatives, friends, and sometimes bosses to cover the application costs. This charge is deducted from their down payment if the applicant is given the opportunity to buy a TOKİ house; otherwise, it is refunded. TOKİ runs the application process in partnership with local municipalities. Applicants are required to pay the application fee to TOKİ's accounts in local public bank branches and submit the bank receipt in the application together with other official documents. For the low-income group application, the major requirements are proof of total household income (below 6,000 TRY for Istanbul in the year 2019), place of residence (Istanbul), age (over 25), Turkish citizenship, and not owning a house anywhere else (including the spouse and children) (TOKİ, 2020c). People begin queuing in front of the municipality and banks days in advance; as Ahmet describes, "I waited in the line from the crack of dawn to evening adhan for 3 days. My boss finally said, 'do not take the trouble of coming to work if you miss another day'. It's tiring, you know; you drop everything to be a homeowner" (48, worker at a bakery).

TOKİ's housing provision is established as campaign, in which homeownership is praised as the only road to upward mobility and is turned into a spectacle of lottery draws for housing units, where people assemble in hopeful crowds. Especially in Istanbul, people look forward to the next housing campaign and the announcements are followed by eager promotion on media channels. While the application process requires individuals to prove that they are in need of housing and cannot afford it in the market, the lottery process makes them feel chosen to qualify for homeownership. Yet, only a small minority gets the chance to become homeowners. Hediye's story is one of several related ones that residents told me about how they felt "seen" and "gifted":

"My husband was a janitor back then. Since we came to Istanbul 25 years ago, we've been living in janitor units, in the basements. Every day I was praying for my own house, though we didn't have the means to afford it. My neighbour was working in the (local) municipality. One day, god bless her, she told me about TOKİ houses to be built. She helped us to fill out the forms. Then, we borrowed the application fee (1,000 TRY) from my brother-in-law. When the lottery day came, we went to a big hall jam-packed, like doomsday. We looked at the people and then looked at ourselves. My husband said, 'Hediye, we have no chance in this crowd, let's leave'. I insisted on staying. After a while, they called my husband's name, I couldn't believe my ears. We both cried with joy. The cameras were on us, we were on TV for the next three days (laughs). You know, my god smiled on me, gave me a unit on the 11th floor. From where to where (gestures her hand, meaning from bottom to top)".

(60, cleaner)

In recent years, instead of a gradual long-term housing provision policy, TOKİ began announcing housing projects as yearly campaigns and launching them in mass gatherings such as President and Justice and Development Party (AKP) leader Erdoğan's political rallies. The 2019 Social Housing Campaign for 50,000 housing units was launched by President Erdoğan in March 2019, just before the municipal elections (TOKİ, 2019b). For 50,000 units to be built across Turkey, TOKİ received 636,949 applications (TOKİ, 2019c). A year later, during the fieldwork for this study, TOKİ announced the 2020 Social Housing Campaign, for 100,000 units across Turkey, with 10,000 of those allocated to Istanbul (TOKİ, 2020b). In Istanbul, 10,000 housing units were distributed among three districts (Başakşehir, Arnavutköy, and Tuzla), all of which have been governed by AKP municipalities for the last three election cycles. The 2020 campaign sparked a wave of excitement and enthusiastic conversations among the TOKİ community. The housing estate management's phones kept ringing as people from all over the city called to ask about the eligibility requirements. The scale of the population that these campaigns mobilize is worth critical attention, especially in Istanbul, since it is a matter of not only providing housing for citizens but also consolidating the governing party's (AKP) power in localities through municipalities (Çavuşoğlu and Strutz, 2014). Therefore, bringing a TOKİ project to their district is regarded as a prestigious service by the local leaders.

Moreover, although the campaigns turn into a spectacle, occupying the prime-time news, and keeping the public's everyday agenda busy for weeks, the actual impact of the housing provision in terms of providing adequate housing for low-income households in cities is questionable. For example, for the 10,000 housing units allocated for Istanbul in the 2020 campaign, there were 360,647 eligible applications[5] (TOKİ, 2020b). That means, from a social policy perspective, the program will only serve 2.7% of the low-income households who actually qualify as in need of housing and cannot afford to get one in private market conditions. As a result, the spectacle of the state's housing mobilization obscures the growing housing need and lack of affordable housing options for low-income groups across cities.

Loyal Clients as Financial Subjects: Bearing Debt Burdens

Though TOKİ houses are marketed as affordable options for low-income homebuyers, keeping up with the housing debt and homeownership-related costs while sustaining life in TOKİ is far from being affordable for many. The major challenges for a low-income TOKİ homebuyer, according to the public bank employees responsible for TOKİ accounts, are the financial responsibilities (down payments and monthly instalments) that come with the mortgage debt and legal obligations of homeownership. Although TOKİ projects were claimed "to create a monthly mortgage payment as close as possible to the amount of rent the homeowner could normally afford to pay" (TOKİ, 2011: 43), the administration seems to overlook to inform the homebuyers clearly about the initial costs that they should have prepared for. These payments include a 12% down payment for low-income group housing during the construction period, as well as other expenses such as turnkey fees and utility instalment costs when households move in. Many research participants said that the first couple of years in TOKİ estates were the most financially stressful period of their TOKİ life, during which they stretched themselves too thin to make ends meet. Also, the majority of the households who signed contracts with TOKİ were unaware that they were expected to pay for home insurance and property taxes twice a year, plus stamp tax in addition to mortgage debt instalments. Participants said that they learned about these types of extra fees either after signing the contracts or later when they moved in; therefore, they had to borrow from relatives or sell their valuables to cover those expenses. The provision process reveals that TOKİ low-income housing projects are far from being social housing but rather aim to provide subsidized owner-occupation. Moreover, low-income households' experiences highlight that they are also not as affordable as they are claimed to be regarding the initial costs.

The housing unit's initial price varies depending on the floor and facade of the unit as well as its proximity to social amenities. As a result, even within the same building block, every household pays a different monthly instalment. For instance, among the interviewed households, who are living in the same building, the initial instalments in 2010 ranged from 356 to 450 TRY. Moreover, every six months, their outstanding debt is re-calculated using the period's lowest of the wholesale price index, consumer price index, or civil servant wage increase rate (a price index system that is both unstable and unpredictable). That means their instalments increase every six months resulting in some households paying as little as 655 TRY and some others paying as much as 945 TRY per month for their housing unit in Fall 2019. The uncertainty of what their next instalment will be put TOKİ households in constant anxiety. Tufan (42, factory worker) explains it this way: "I know how much will go into my pocket each month, but I don't know how much will come out. My instalment increases every 6 months but not my wage". Also, households complain that their overall debt never decreases despite the fact that they have been paying their instalments for nearly ten years. The housing unit purchase agreement with TOKİ and the public bank specifies the purchase price of the housing unit but not the cumulative amount they would have paid at the end of the mortgage term (including the interest payments). For the vast majority of households, TOKİ agreement is their first interaction with a financial institution. Therefore, the institutionalization and financialization of low-income housing without necessary financial literacy support leaves households confused and overwhelmed by never-ending debt. When this topic was discussed with a public bank officer responsible for TOKİ accounts, they told

"This is the biggest challenge we have with low-income clients from TOKİ. This is not a standard mortgage debt; their instalments increase twice a year in January and July. They also have to pay the insurance and estate tax. It's written on the contract, but they can't get their heads around it. They think they will pay the same 500 TRY (instalment) for 15 years. Since they get the house from the state, they think of it as a charity. They aren't aware of the fact that it's actually credit. At the same time, we understand it, some families are in dire straits".

(Bank officer, Istanbul)

According to the purchase agreement, if the clients are not able to pay their monthly instalments two times in a row, they get noticed and face the risk of confiscation and repossession (TOKİ, n.d.b). As explained earlier in the chapter, public bank credits for TOKİ low-income groups are different from the standard mortgage credit that commercial banks provide. It is a highly insecure system for households since there is no mortgage insurance. Moreover, until the debt is fully paid off, the housing title deeds remain with TOKİ and the households cannot sell or lease their property. In TOKİ's official website, this is justified as "the fact that the title deeds are not issued until full repayment of the debts minimizes the default of payment of instalments" (TOKİ, n.d.a). That means the continuation of this social provision model appears to be guaranteed by coercive measures, such as contract termination and confiscation in the case of default payments, which shift the entire financial risk on to homebuyers.

Moreover, interviews with TOKİ officials uncover some discrepancies between the rule and the practice. The housing provision model for low-income groups is promoted as social housing, and during the application process, households are asked to prove that they *need* these houses, not that they have the financial means to afford them. However, the housing finance model is designed to expect households to have regular incomes and financial solvency, in other words, ability to meet their debts. Accordingly, when TOKİ officials were asked how TOKİ justifies contract terminations in social housing in the case of default payments, they explained that "although it says in the contract to send a notice after two default payments", they "do not do it until the 8th non-payment" (TOKİ officer, Ankara). They continue

> "Those who have specific problems come to our office to explain themselves. Some say they got divorced, some say they had an accident. We deal with all of them, do not send anyone back crying. They are low-income people, maybe they don't have any bread to eat. But if they promise to pay in 3 months, they pay in 3 months. As long as they speak decently and truthfully, we trust their word, we help them out. At the end of the day, these are minuscule numbers for TOKİ and don't affect our operational capability. [...] As we run this logic, our rate of collection is as high as 93 percent.
> Interviewer: If you'd stuck to the agreement (sending notices), what would that rate be?
> If we did it after 2 default payments, maybe this rate (of default payments) would rise to 20 percent".

> (TOKİ officer, Ankara)

According to TOKİ's published housing provision figures, a 20% rate of default payments corresponds to approximately 60,000 households (TOKİ, 2019a). This means that one in five households qualifying for social housing will not usually be able to afford to live in TOKİ, despite the official discourse of affordability. Confiscations, on the other hand, are relatively few, according to TOKİ authorities. However, the disparity between the rule and the practice provides space for debt negotiations between the TOKİ authorities and debtors. Homebuyers appear to accept that they will be fine as long as they are grateful, and the authorities appear to acknowledge that the policy seems to be working as long as such shortcomings are overlooked. Justified as showing sympathy and providing convenience, nevertheless this approach allows veiling the flaws and vulnerabilities within the system. As a result, initiative leads to how Lorey (2015) describes as *governmental precarization*, in which insecurity is instrumentalized, in this context rendering TOKİ debtors governable and complicit.

The Promise of Profit: Navigating the Quasi Real-Estate Market

TOKİ estates have seen renting, subletting, and different forms of property transfers in recent years, while a quasi-real-estate market is emerging through word of mouth among real estate agents, housing estate management, and households. On the research site, TOKİ housing units currently have a certain financial value, yet continuing debt instalments and the binding rules of the purchase agreement still keep some households off the real estate market. Residents are motivated by the prospect of one day turning their house into an asset, especially in an economy where the tenancy is viewed as precarity. In this type of system, exchange value prevails use-value, paving the steps towards the integration of low-income groups into the real estate market as homeowner classes.

The analysis of housing unit occupancy data at the TOKİ housing estate management office shows that by December 2019, 30% of the housing units were occupied by tenants while 19.4% were occupied by second or third owners. Just 259 of the 562 housing units were still owned by the original owners who signed the purchase agreements with TOKİ. This means that, since the lots were drawn in 2010, almost half of the housing units in this low-income housing estate have been absorbed by the local real estate market, either leased or sold out. Since the purchase agreements for this estate have a 15-year period, one can presume that the original debtors had either paid off their entire debt earlier (via TOKİ's debt clearance campaigns) or let the new buyer pay off their debts and transferred their ownership rights. It is difficult to determine if any case of ownership transfer violates the purchase agreement or not but selling or leasing out a flat is a constant topic of discussion in the estate at coffeehouses, estate management offices, and among shopkeepers.

The market value of TOKİ houses on this research site has tripled over the last ten years. Homebuyers signed contracts starting at 80,000 TRY in 2010, but due to rising interest rates and instalment payments, they have paid more over the years. Even so, their house prices have risen significantly over the last decade as the demand for affordable housing has outstripped the supply, especially in Istanbul. In fall 2019, the average price of a TOKİ low-income housing unit in the study area was 200,000 TRY (online research and interviews with estate agents). As of the writing of this chapter in fall 2020, the average price of the same unit had increased to 250,000 TRY. In the case of rental units, housing estate management clarified that since homeowners can charge rents (up to 1,000 TRY) that are higher than their monthly mortgage payments (700–900 TRY), they choose to lease their houses and go to live in some other parts of the city where rents are cheaper, or they share housing with the extended family.

Since 2009, there have been high-income residential projects being built by private contractors around the site. These luxury housing developments tend to be pushing up overall housing prices in the neighbourhood. There are five different real estate agents located around the research site. During the interviews, estate agents stated that in a single month, they post approximately 80 new properties (for sale or lease) on real estate websites, and they normally sell or lease all of them in that same month. Their portfolios contain properties both from the luxury housing estates and TOKİ low-income housing estates. They stated that "it has been a fast-moving market with strong demand and competition". Some agents complained that the "profile of the people living in TOKİ estates impacted the rising real-estate image of the area negatively". While some others suggested that "these TOKİ estates should be redeveloped as high-income residences (via UTPs) to fit the luxury developments of the neighbourhood". On the real estate listing sites, TOKİ houses are advertised as being 5 minutes driving distance away from the marina, international airport and major highways with the goal of attracting a different social class to the area. A TOKİ resident Ayşe tells

"When we see those advertisements, my neighbours and I laugh. They say 'it is 5 minutes to the marina, 5 minutes to the airport'...what good does that do me? My son is a depot worker at the airport. The other day he missed his work service. There are no public buses to the airport from here. So, he had to take a taxi to work so as not to lose his job; and paid 30 TRY, worth of his half day's work!"
(44, temporary cleaning worker at the municipality)

Yet, for many other participant households, rising property values are beneficial for their situation. On the one hand, they told about the financial hardships they have experienced, struggling to get by with debts in TOKİ, one resident describing as "similar to you doing a doctorate on TOKİ my friend, I have a professorship on getting by" (Ali, 38, factory worker). On the other hand, they spoke about their plans to buy a bigger house somewhere else once they paid off their debts and sold their TOKİ house. Large-scale investments (such as the marina or new highway) around the district, as well as luxury residences being built closer to their estate, seem to be perceived as opportunities. Another TOKİ resident (Mehmet, 47, factory worker) was planning to sell his house but was waiting for the right time to get a good offer. He still had five years of debt payment, but he felt they could work it out with the new buyer. He portrayed the research district as "rising, as the new Yeşilköy of the Anatolian side" (Yeşilköy is a district on the European side of Istanbul, nearby Atatürk Airport, with very high property values) and advised me to buy a house in this TOKİ estate if I had some savings. Another resident summarized the dilemma they were in as

"We are poor people. Life does not give us the freedom to choose where we want to live. We fell into this TOKİ project, you know, our life here is like a kind of conviction. But, the value of my house has tripled over the last ten years, maybe it's worth the prison life after all (laughs)".
(Bekir, 45, dockworker)

Bekir's words reflect how households bear the burdens of the present in exchange for the potential profits of the future, constantly weighing their options. Rising property values lead to consent to debt-based housing provision, forced placements, increasing household indebtedness, and the control of the state over the affordable housing market. The victory of exchange value in TOKİ estates mobilizes TOKİ households around negotiating with the authorities, bargaining with the market actors, and profiting from the ongoing neoliberal transformation of the city. State-led housing provision processes coupled with the inclusion of masses into institutionalized housing finance turn low-income households into financial subjects and gradually savvy actors of the emerging real estate market in TOKİ.

Conclusions

This chapter discusses the role of indebtedness to the state on the conduct of citizens by examining the debt-based housing provision and everyday debt relations in a low-income TOKİ housing estate in Istanbul. The chapter theorizes negotiations of finance and debt in three interconnected categories: (1) provision, (2) debt, and (3) profit. The chapter argues that in TOKİ's social housing programs for low-income groups, access to housing is conditioned upon homeownership through long-term financial obligations. As a result, needs-based eligibility in the application without necessary financial means and literacy to manage debts in the long-run leaves low-income households at the mercy of authorities. Moreover, while being indebted to the state for their house, TOKİ residents also become financial subjects via state-owned banks' credit mechanisms, payment plans, and additional financial products. In a financialized real estate environment created by TOKİ and its local partners, households understand that they were made to buy houses that are somewhat gaining value; and the prospect of profiting in real estate legitimizes debt-based housing provision, rampant household indebtedness, and the neoliberal transformation of urban land. As a result, debt-based housing provision is less enacted by rules and regulations but more manifested through invisible and quotidian negotiations in several scales among multiple actors as policy plays in the field.

Though the research findings might be specific to a particular housing provision mechanism and the particular cultural context of TOKİ housing communities in Turkey, they speak to theoretical questions about labour, class, and citizenship that are being addressed through social sciences. This chapter turns to scholarly contributions by offering an analysis of lived experiences of housing debts that connects TOKİ households to neoliberal housing policies and state-managed housing finance as financial subjects. The results provide fresh insights into how neoliberal housing policies rely upon and rework local communities as well as networks of labour and politics. The practical contribution will be a nuanced understanding of the politics of (mortgage) debts to the state, providing insight for urban advocates to counter the growing inequalities in urban housing policy and practice.

Notes

1 A TOKİ low-income mass housing estate is a group of multi-storey, mass-produced building blocks that are occasionally built on urban public land in partnership with local municipalities. These dense and repetitive housing projects can comprise hundreds to thousands of housing units and, in many cases, serve as dormitory settlements at the outskirts of cities.

2 Calculated by the author using data from several sources including Ministry of Family, Labour, and Social Services (minimum wages by years), The Central Bank of the Republic of Turkey (total household debt in TRY, in years 2002 and 2020), Economics Institution of Turkey (total number of households in 2002), and Turkish Statistical Institute (total number of households in 2020).

3 There are three separate housing estate management offices in this low-income TOKİ housing site that are responsible for different building block groups. This is the TOKİ housing estate's management unit, where the author conducted their ethnographic research. The estate is governed by a board of directors, who are elected members from each of the estate's building blocks. Aside from the board, the management office tasks are run by an office manager, an accountant and a group of janitor-gardeners who are responsible for the maintenance. The monthly maintenance fee that all estate residents are obliged to pay for the housing estate to be cared for is collected by the management office.

4 The term "muhtar" in Turkish refers to the local chief. Muhtar is an officially elected neighbourhood representative. Muhtars are elected for five-year terms in Turkey by local elections.

5 Derived by the author from the qualifying applicants lists that are available on TOKİ official website through the Application and Ownership Right Draw Results Database. Available at: http://talep.toki.gov.tr/index.html

References

Aalbers MB (2016) *The Financialization of Housing: A Political Economy Approach.* London: Routledge.

AÇSH (2021) *Daily and Monthly Minimum Wages by Years.* Aile, Çalışma ve Sosyal Hizmetler Bakanlığı (Ministry of Family, Labour, and Social Services). Available at: https://www.ailevecalisma.gov.tr/tr-tr/asgari-ucret/asgari-ucretin-net-hesabi-ve-is-verene-maliyeti/. Accessed 3 March 2021.

Aslan S, Güzey Ö (2015) Affordable housing provision: A case study of TOKİ Ankara Kusunlar low-income housing. *Journal of Ankara Studies* 3(1): 42–53.

Candan AB, Kolluoğlu B (2008) Emerging spaces of neoliberalism: A gated town and a public housing project in İstanbul. *New Perspectives on Turkey* 39: 5–46.

Çavuşoğlu E, Strutz J (2014) Producing force and consent: Urban transformation and corporatism in Turkey. *City* 18(2): 134–148.

CBRT (2006) *Financial Stability Report: 2, June 2006.* Ankara: The Central Bank of the Republic of Turkey, 12–17. Available at: https://www.tcmb.gov.tr/wps/wcm/connect/TR/TCMB+TR/Main+Menu/Yayinlar/Raporlar/Finansal+Istikrar+Raporu/. Accessed 9 March 2021.

CBRT (2020a) *Financial Stability Report: 31, November 2020.* Ankara: The Central Bank of the Republic of Turkey, 25. Available at: https://www.tcmb.gov.tr/wps/wcm/connect/TR/TCMB+TR/Main+Menu/Yayinlar/Raporlar/Finansal+Istikrar+Raporu/. Accessed 9 March 2021.

CBRT (2020b) *Financial Calculations Report, Third Quarter.* Ankara: The Central Bank of the Republic of Turkey, 1–13. Available at: https://www.tcmb.gov.tr/wps/wcm/connect/TR/TCMB+TR/Main+Menu/Yayinlar/Raporlar/Finansal+Hesaplar+Raporu/. Accessed 9 March 2021.

Di Feliciantonio C (2016) Subjectification in times of indebtedness and neoliberal/austerity urbanism. *Antipode* 48(5): 1206–1227.

Di Muzio T, Robbins RH (2016) *Debt as Power*. Manchester: Manchester University Press.

Doğru HE (2016) The "benevolent hand" of the Turkish State: Mass Housing Administration, state restructuring and capital accumulation in Turkey. Unpublished doctoral dissertation. Toronto: York University.

Ergüder B (2015) 2000'li yıllarda Türkiye'de hanehalkı borçlanması: Konut kredileri ve toplumsal refah. Praksis 28: 99-127.

Erman T (2016) *"Mış gibi site": Ankara'da bir TOKİ-gecekondu dönüşüm sitesi.* Ankara: İletişim.

Erol I (2019) New geographies of residential capitalism: Financialization of the Turkish housing market since the early 2000s. *International Journal of Urban and Regional Research* 43(4): 724–740.

Federici S (2014) From commoning to debt: Financialization, microcredit, and the changing architecture of capital accumulation. *South Atlantic Quarterly* 113(2): 231–244.

Few R (2002) Researching actor power: Analyzing mechanisms of interaction in negotiations over space. *Area* 34: 29–38.

García-Lamarca M, Kaika M (2016) "Mortgaged lives": The biopolitics of debt and housing financialization. *Transactions* 41(3): 313–327.

Graeber D (2011) *Debt: The First 5,000 Years*. New York: Melville House Printing.

Hall SM (2016) Everyday family experiences of the financial crisis: Getting by in the recent economic recession. *Journal of Economic Geography* 16(2): 305–330.

Han C (2012) *Life in Debt: Times of Care and Violence in Neoliberal Chile*. Berkeley: University of California Press.

Harker C (2017) Debt space: Topologies, ecologies and Ramallah, Palestine. *Environment and Planning D: Society and Space* 35(4): 600–619.

Hatiboğlu Eren B (2017) *İmkansız Medeniyet: TOKİ-Aktaş Mahallesi Örneğinde Feminist Bir Etnografi*. Ankara: Ayizi Kitap.

Joseph M (2014) *Debt to Society: Accounting for Life under Capitalism*. Minneapolis: University of Minnesota Press.

Karaagac EA (2020) The financialization of everyday life: Caring for debts. *Geography Compass* 14(11): e12541.

Karaçimen E (2014) Financialization in Turkey: The case of consumer debt. *Journal of Balkan and Near Eastern Studies* 16(2): 161–180.

Kear M (2013) Governing Homo Subprimicus: Beyond financial citizenship, exclusion, and rights. *Antipode* 45(4): 926–946.

Kuyucu T (2014) Law, property and ambiguity: The uses and abuses of legal ambiguity in remaking Istanbul's informal settlements. *International Journal of Urban and Regional Research* 38(2): 609–627.

Kuyucu T, Ünsal O (2010) Urban transformation as state-led property transfer: An analysis of two cases of urban renewal in Istanbul. *Urban Studies* 47(7): 1479–1499.

Lazzarato M (2012) *The Making of Indebted Man: An Essay on the Neoliberal Condition*. Cambridge: MIT Press.

Lazzarato M (2015) *Governing by Debt*. Semiotex(e) Intervention Series 17. Cambridge: MIT Press.

Lelandais GE (2015) Urbanisation under neoliberal conservatism in Turkey. *Research Turkey, Centre for Policy and Research on Turkey* 4(7): 54–67.

Lorey I (2015) *State of Insecurity: Government of the Precarious*. New York: Verso.

Lovering J, Türkmen H (2011) Bulldozer neo-liberalism in Istanbul: The state-led construction of property markets, and the displacement of the urban poor. *International Planning Studies* 16(1): 73–96.

Martin A (2015) *Right to buy or the right to lifelong debt?* [Blog post] The New Economic Foundation. Available at: http://www.filmsforaction.org/articles/right-to-buy-or-the-right-to-lifelong-debt/. Accessed 12 March 2021.

Montgomerie J (2013) America's debt safety-net. *Public Administration* 91(4): 871–888.

Montgomerie J, Tepe-Belfrage D (2017) Caring for debts: How the household economy exposes the limits of financialization. *Critical Sociology* 43(4–5): 653–668.

Özdemir D (2011) The role of the public sector in the provision of housing supply in Turkey, 1950–2009. *International Journal of Urban and Regional Research* 35(6): 1099–1117.

Özdemir Sarı ÖB, Aksoy Khurami E (2018) Housing affordability trends and challenges in the Turkish case. *Journal of Housing and the Built Environment* 1–20. DOI: 10.1007/s10901-018-9617-2.

Peters MA (2016) Biopolitical economies of debt. *Analysis and Metaphysics* 15: 7–19.

Pınar A, Demir M (2016) The impact of social housing program on the demand for housing in Turkey: A cross-section analysis. *Applied Economics and Finance* 3(4): 20–28.

Rolnik R (2013) Late neoliberalism: The financialization of homeownership and housing rights. *International Journal of Urban and Regional Research* 37(3): 1058–1066.

Ronald R, Elsinga M (2012) *Beyond Home Ownership: Housing, Welfare and Society.* Oxon: Routledge.

Sarıoğlu-Erdoğdu GP (2014) Housing development and policy change: What has changed in Turkey in the last decade in the owner-occupied and rented sectors? *Journal of Housing and the Built Environment* 29(1): 155–175.

Sharp JP, Routledge P, Philo C et al. (2000) Entanglements of power: Geographies of domination/resistance. In: Sharp JP, Routledge P, Philo C et al. (eds) *Entanglements of Power: Geographies of Domination/Resistance.* New York: Routledge, 1–42.

Soederberg S (2014) *Debtfare States and the Poverty Industry: Money, Discipline and the Surplus Population.* New York: Routledge.

Tansel CB (ed.) (2017) *States of Discipline: Authoritarian Neoliberalism and the Contested Reproduction of Capitalist Order.* London and New York: Rowman & Littlefield.

TBB (2020) November 2020 Monthly Bulletin. The Banks Association of Turkey, Risk Centre. Available at: https://www.riskmerkezi.org/Content/Upload/istatistikiraporlar/ekler/2626/Risk_Merkezi_Aylik_Bulten_Ozeti_Kasim_2020.pdf. Accessed 12 March 2021.

TEK (2003) *2002 Household budget survey: Income distribution and consumption related to expenditures evaluation of the results*, Discussion Paper. Türkiye Ekonomi Kurumu. Available at: http://www.tek.org.tr/dosyalar/HANE2002.pdf. Accessed 12 March 2021.

TOKİ (2011) *TOKİ Corporate Profile 2010/2011.* Ankara: Housing Development Administration of Turkey. Available at: https://www.TOKİ.gov.tr/yayinlar. Accessed 10 March 2021.

TOKİ (2019a) *TOKİ Corporate Profile Document.* Ankara: Housing Development Administration of Turkey. Available at: https://www.TOKİ.gov.tr/yayinlar. Accessed 10 March 2021.

TOKİ (2019b) 50 Thousand new social housing units from TOKİ. TOKİ Official Website. Available at: https://www.TOKİ.gov.tr/haber/TOKİden-50-bin-yeni-sosyal-konut. Accessed 3 March 2021.

TOKİ (2019c) Lot drawing has started for 50 thousand housing units. TOKİ Official Website. Available at: https://www.TOKİ.gov.tr/haber/50-bin-yeni-sosyal-konut-icin-kuralar-basladi. Accessed 3 March 2021.

TOKİ (2020a) Summary of operations. TOKİ Official Website. Available at: https://www.TOKİ.gov.tr/faaliyet-ozeti. Accessed 3 March 2021.

TOKİ (2020b) 100 Thousand social housing units campaign—Application and ownership right draw results information database. TOKİ Official Website. Available at: http://talep.TOKİ.gov.tr/. Accessed 3 March 2021.

TOKİ (2020c) 100 Thousand social housing project application information form. TOKİ Official Website. Available at: https://i.toki.gov.tr/100BinKonut/. Accessed 3 March 2021.

TOKİ (n.d.a) Housing programs. TOKİ Official Website. Available at: https://www.TOKİ.gov.tr/en/housing-programs.html. Accessed 3 March 2021.

TOKİ (n.d.b) TOKİ—Mass housing purchase agreement. Housing Development Administration of Turkey. Obtained from TOKİ Administration, Department of Housing Credits, Ankara (in Turkish).

TURKSTAT (2020) Number of households with respect to type and size. In: Turkish Statistical Institute *Adrese Dayalı Nüfus Kayıt Sistemi Sonuçları, 2020*, Ankara. Available at: https://data.tuik.gov.tr/Bulten/ Index?p=Adrese-Dayali-Nufus-Kayit-Sistemi-Sonuclari-2020-37210. Accessed 11 March 2021.

Türkün A (2011) Urban regeneration and hegemonic power relationships. *International Planning Studies* 16(1): 61–72.

Türkün A (ed) (2014) *Mülk, Mahal, İnsan: İstanbul'da Kentsel Dönüşüm*. Istanbul: Bilgi Üniversitesi Yayınları.

Türkün A (2015) Ruthless transformation efforts in the housing areas of the urban poor and implications for the right to housing. In: Research Committee on Urban and Regional Development *RC21 International Conference on the Ideal City: Between myth and reality*. Urbino, Italy, 45.

Ünsal BO (2015) State-led urban regeneration in Istanbul: Power struggles between interest groups and poor communities. *Housing Studies* 30(8): 1299–1316.

Whitehead C (2017) Social housing models: Past and future. *Critical Housing Analysis* 4(1): 11–20.

Part III
Housing Market Experiences

9 Understanding Residential Vacancy and Its Dynamics in Turkish Cities

H. Kübra Kıvrak and Ö. Burcu Özdemir Sarı

Introduction

Turkey has a high-performing housing market in terms of new construction. Housing production has far exceeded the requirement over the last two decades (Türel and Koç, 2015; Özdemir Sarı, 2019). Residential construction by the private sector and the public sector's involvement in housing production as a direct actor have been deliberately encouraged since the early 2000s. In this period, increasing the performance of the construction sector is seen as a way to ensure economic growth and reduce short-term unemployment. Following the 2002 elections, first a 'countrywide housing programme' and then an 'urban transformation programme' were initiated by the Justice and Development Party (AKP) governments (Özdemir Sarı, 2013). In this way, new housebuilding was supported on vacant land allocated for development and in the existing residential areas, squatter housing areas, and historical parts of cities through redevelopment as part of urban transformation. As a result of this approach, a significant housing surplus emerged across the country, leading to high residential vacancy rates. However, the geographical distribution of the housing output in the country was not even. Construction activities intensified in some housing markets, whereas other markets suffered from a lack of investment. As a result of this geographically uneven housing output, historically existing housing disparities across the country have deepened over the last 20 years. Today, both a housing shortage and excess housing production exist simultaneously (Özdemir Sarı, 2019).

The first time that annual housing production exceeded 1 million units in the country was in 2014. This was followed by the production of 1.4 million dwelling units in 2017. During these years, 500–550,000 households were added to the number of households annually. While the annual average housing production during 2000–2010 was 450,000, this figure almost doubled in the period 2011–2020 to an average of 810,000 units despite the 2018 economic crisis and the Covid-19 pandemic, which negatively affected housing production. As a result, an oversupply of housing emerged in Turkey due to the continuous support given by governments to the construction sector and speculative profit opportunities foreseen by the construction firms.

DOI: 10.4324/9781003173670-12

Furthermore, house prices continued to increase until 2017 despite the excess housing production in the country. In the third quarter of 2017, just after a warning by the International Monetary Fund (IMF) about the possibility of an emerging housing bubble in Turkey, real house prices started to decline (Özdemir Sarı, 2019). The year 2018 marked the beginning of a difficult period for the housing sector. The economic crisis had a negative effect on the activity of construction firms. Housing output and housing transactions (particularly mortgaged sales) declined in 2018 and 2019. During the same period, construction companies experienced many shutdowns, while a decline was observed in the number of new companies entering the sector. In this context, the oversupply of housing and vacancy rates have become a major issue in the country. The government has implemented various measures to increase the housing transactions involving the newly built stock, especially through mortgages. In the course of these discussions, it also became clear that the amount of surplus housing stock in the country is unknown. Thus, the relevant public agencies started to prepare an inventory of the housing stock.

In this context, the present study makes a countrywide assessment of the vacancy problem, employing official statistical data related to households and housing stock. It is argued that neither the vacancy rate nor the reason for the vacancy is uniform across the housing market. Accordingly, different residential vacancy types are examined through a case study to provide inputs for policy and planning. The study is organized as follows. The next section reviews the basic theoretical and empirical knowledge on residential vacancy. Then an assessment of the countrywide residential vacancy problem is made and current policies and measures developed to cope with it are examined. A case study is presented and neighbourhood-level insights are provided about the dynamics of the residential vacancy in the Gemlik District of Bursa Province to elaborate an understanding of the vacancy types. Conclusions are presented in the final section.

Residential Vacancy: Theoretical and Empirical Background

Durability and immobility are among the main features that distinguish housing from other commodities. The durability of housing implies that due to the long physical lifespan of residential buildings, a decline in the demand for a particular location or housing type will lead to vacancy (Molloy, 2016). Sometimes vacancies emerge not due to demand-side changes but directly due to supply-side actions such as the rate of new construction (Belsky and Goodman, 1996). Since housing is spatially fixed on land, it is not possible to transfer the vacant stock to another location where there is an existing demand for housing. Moreover, an increase in demand may lead to vacant stock being absorbed since the housing supply is slow to respond to demand changes. The absorption, in this case, is only possible if available vacant stock is a close substitute for the demanded housing units in terms of location and characteristics. Moreover, absorption may occur due to the demolition of a residential building or its conversion to non-residential use (Magnusson Turner, 2012). In this context, it is possible to define vacancy as a supply–demand imbalance in a housing market, and it is argued in the literature that, even in equilibrium, there is always some residual housing (Vandell, 2012).

In the residential vacancy literature, problems related to the amount of vacancy are usually evaluated based on vacancy rates and the duration of the vacancies (see, for instance, Gabriel and Nothaft, 2001; Wood et al., 2006; Huuhka, 2016; Molloy, 2016). In general, the vacancy rate is defined as the share of unoccupied housing units in the total housing stock in a market (or a geographic area). Here, the definition of the vacant units becomes significant since it varies between the contexts of different countries. For instance, in Switzerland, vacant dwellings refer to flats or single-family houses that are fit for habitation, unoccupied, and available in the market for rent (Thalmann, 2012). As Vakili-Zad and Hoekstra (2011) note, in Greece, vacant dwellings are empty dwellings with at least one habitable room in permanent and independent structures, and, in Italy, dwellings that have no permanent occupants are considered vacant. In the Portuguese context, vacant dwellings that are kept out of the market are not counted (Vakili-Zad and Hoekstra, 2011).

Vacancy rates could be expressed as a rental vacancy, sale vacancy, or overall vacancy covering both types. A certain amount of vacant dwelling units is considered necessary to allow households' mobility (Glock and Häussermann, 2004; Magnusson, 2012; Couch and Cocks, 2013) and accommodate newly formed households (Gu and Asami, 2016). In the literature, this kind of vacancy is usually referred to as the natural, equilibrium, or optimal vacancy rate; yet there is no consensus about its level (Thalmann, 2012). As Rosen and Smith (1983) highlight, the optimal vacancy rate in any market refers to a rate at which there is no excess demand or excess supply. Varying definitions of vacant housing and different housing systems in different country contexts are just two of the many reasons behind the lack of consensus on the optimal vacancy rate. Belsky (1992), in the US context, criticizes the common practice of accepting a 5% rental vacancy rate as equilibrium vacancy and argues that in equilibrium the rental vacancy rate varies among areas and types of units. Thalmann (2012) reports that the equilibrium vacancy rate in the Swiss housing market is remarkably low, 1–1.5%, where vacancy is defined as unoccupied dwellings for rent or sale. Glock and Häussermann (2004), on the other hand, accept rates of 3–5% as the natural vacancy rate for a well-functioning housing market. Couch and Cocks (2013) assume that best-performing housing markets in the country may proxy for the frictional (short-term) vacancy and report that a 2.5% vacancy rate could be accepted as frictional vacancy.

Vacancy rates fluctuate in time and differ between markets and submarkets (Belsky and Goodman, 1996; Gabriel and Nothaft, 2001; Hagen and

Hansen, 2010; Huuhka, 2016). A comparison of the optimal vacancy rate and the actual one in a housing market at a point in time gives clues about the possible rent/price changes and the extent of shortage/excess production. According to Belsky and Goodman (1996), the incidence of vacancy is affected by mobility triggers (e.g., demographic conditions and local economic conditions), the rate of new construction, and rent control laws. The average duration of the vacancies, on the other hand, is influenced by the cost and quality of market information, the size and heterogeneity of the market, the proportion of ill-matched units to current demand, and the extent of professional management (Belsky and Goodman, 1996). Although the duration of the vacancy is a very significant variable for residential vacancy studies, it is not covered by most of the datasets providing information on vacant housing (Molloy, 2016).

Residential vacancy rates are of great importance for public policy and urban planning since they represent several problems in the housing market that policy and planning focus on. Actual vacancy rates below the optimal vacancy rate in a market indicate housing shortages, and those above the optimal rate signify an oversupply of housing. In the case of shortage, overcrowded living and affordability problems are likely to be observed. Furthermore, particularly in developing countries, shortage triggers unauthorized housing construction (Özdemir Sarı, 2019). In addition, high vacancy rates indicate multi-faceted resource inefficiency at regional and urban scale: (i) waste of resources invested in housing (Couch and Cocks, 2013; Han, 2014; Molloy, 2016), (ii) waste of resources invested in infrastructure (Gu and Asami, 2016), and (iii) unjustifiable development leading to urban spatial expansion (Özdemir Sarı, 2015). When considered at the neighbourhood scale, the authors of several studies argue that vacant homes create problems such as crime and declining quality and property values (Couch and Cocks, 2013; Han, 2014; Gu and Asami, 2016; Huuhka, 2016; Molloy, 2016; Jensen, 2017). At the housing unit scale, vacant houses represent unpaid repair and maintenance costs. As Molloy (2016) argues, houses that experience a long vacancy period are less likely to receive adequate maintenance; thus, renovation is necessary before they can be resided in.

Residential vacancy types may differ in a housing market; thus, response to different categories may require diverse interventions in terms of public policy and urban planning. The first distinction can be made between intentional and unintentional vacancies, that is, whether the vacant dwelling units are purposefully left vacant or not. Hoekstra and Vakili-Zad (2011) note that dwellings bought for investment purposes, dwellings left behind during rural to urban migration, and dwellings as family possessions waiting for a family member to occupy them are often not offered for rent or sale in Spain. Similarly, summer houses, holiday homes, and mountain homes are considered intentionally left vacant. This type of vacancy is still problematic in terms of infrastructure and service provision and safety problems when unoccupied from the policy and planning perspective. Molloy (2016) argues that unintentional vacancy has to be considered in order to understand the extent of the underutilized housing stock since dwelling units that are not intended to be vacant reveal a supply–demand imbalance.

Another distinction can be made regarding the habitability of vacant dwellings. If the vacant houses are abandoned and not suitable for residency, the public policy response would probably be to urge demolition. The unfitness for occupation may be due to functional abandonment (e.g., lack of sealed doors/windows), financial abandonment (e.g., financial responsibilities are not met by the owner), and physical abandonment (e.g., deterioration due to lack of maintenance) (Hillier et al., 2003; Han, 2014). In this sense, abandonment may be observed as an area-based problem or at a single property scale, both of which require different policy responses. Abandoned properties are frequently observed in historical parts of cities and in the context of shrinking cities. Understanding different types of residential vacancies may shed light on public policy and urban planning to cope with and monitor housing market problems related to vacancy. In this context, it is necessary to monitor vacancy rates, the duration of vacancies, the type of vacant stock, the reason for the vacancy, and whether the vacant dwelling units are offered for sale or rent (Aksoy Khurami and Özdemir Sarı, 2020).

High Housing Output and Its Outcomes in Turkey

In Turkey, the housing problem has been considered a quantitative problem for years. Thus, housing policies have continuously targeted increasing the housing supply. With the adoption of construction sector-based economic growth, housing production and construction activities have increased throughout the country. Particularly in the last two decades, an oversupply of housing has begun to be visible to the naked eye. The implications of excess housing production have never been investigated on the policy side. Whether the produced houses are circulating in the market, for which households these houses are affordable, and the effect of this production on house prices on the macro scale remain unexplored. The policies focused only on housing production and managing and monitoring the existing housing stock; the problems experienced in the existing housing areas and the problems related to supply–demand match were ignored, resulting in high vacancy rates. As of 2014, out of 81 provinces, 39 are exhibiting significant levels of surplus housing stock (Özdemir Sarı, 2019).

The Extent of the Residential Vacancies in Turkey

In Turkey, there is no official statistical information on vacancy rates or the duration of vacancies. However, there are several statistical sources that provide a general understanding of surplus housing production, such as

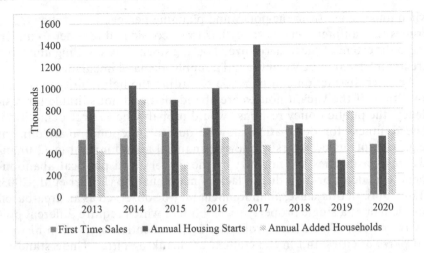

Figure 9.1 Housing production, house sales, and household growth in Turkey during
the period 2013–2020.

Source: Prepared by the authors based on TURKSTAT (2021a, 2021b, 2021c).

Building Permit Statistics, House Sale Statistics, and the National Address
Database (UAVT). Building Permit Statistics provides information about
the authorized part of the housing stock and cover predominantly urban
areas. In other words, these data do not cover squatter housing but do cover
housing units in rural areas if the rural area is within a municipal boundary.
House Sale Statistics, on the other hand, represent all legal transactions in
the country. It is possible to identify first-time sales and second-hand sales
as well as mortgaged and non-mortgage sales. Since 2013, all settlements in
Turkey have been included in House Sale Statistics. Employing these datasets, Figure 9.1 compares annual additions to the stock with first-time sales
and annual additions to households.

Throughout most of the last 20 years, the number of annual housing
starts has been higher than the growth in households. With the impact of the
2018 economic crisis and the pandemic in 2020, housing production is currently lagging behind household growth for the first time in a long period.
During the period 2005–2018, annual housing production never fell below
500,000 units (TURKSTAT, 2021a). As Figure 9.1 shows, housing production was double household growth in 2013 and 2016 and triple in 2015 and
2017. During the period 2013–2020, housing units produced were 2.1 million
above added households. Undeniably, this figure has to be interpreted with
caution since some part of the excess production may meet the backlog need
in the country and some other part may reflect multiple ownership, which is
a widespread phenomenon. Households become multiple owners for several

reasons: having a summer/holiday home, accommodating their children, or just for investment purposes.

On the other hand, a comparison of housing production with first-time sales gives a more accurate picture of the supply–demand imbalance. When first-time sales from the stock are examined in Figure 9.1, except for the 2019 figures, sales from the newly built stock are observed to lag behind annual production levels in all years examined. During the period 2013–2020, only 4.6 million of the 6.6 million new housing units were subject to transactions. However, it must be acknowledged that a small share of new housing production may be unfinished constructions due to various reasons. In other words, approximately 2 million surplus housing stock has been created, and most of it remains in possession of the construction companies. That is the reason vacancy and housing sales have become a central political issue in the country recently.

The UAVT also provides some relevant information for the examination of surplus housing stock, such as the total number of dwelling units at the neighbourhood level. However, due to its aggregation level, although the data include information about the number of regular residential units, housing units for government employees, summer houses, and residential structures under construction, it is not possible to distinguish them within the data. Furthermore, whether these housing units are in poor condition and dilapidated is unknown. In other words, the available data do not allow the identification of intentional/unintentional vacancy and whether the dwelling is fit for occupation. Moreover, these data cover all of the country's housing stock, which means the authorized/unauthorized parts of the stock and rural/urban areas are taken altogether. Previous attempts to understand the residential vacancy dynamics through the UAVT data reveal that inputs obtained from field studies show lower vacancy levels than the estimated vacancy rates from the UAVT data (Özdemir Sarı, 2015; Aksoy Khurami and Özdemir Sarı, 2020). This necessitates the examination of residential vacancy through field studies to understand its dynamics. Still, examination of the UAVT data with respect to households gives another picture of the supply–demand imbalance (Figure 9.2). As seen in Figure 9.2, the number of dwelling units is significantly higher than the number of households during the period 2010–2020. These data include almost every dwelling in the country, such as abandoned properties and summer houses. However, it is a fact that the number of households increased 1.3-fold, whereas the number of dwelling units increased 1.4-fold in the period 2010–2020.

Although it is not possible to determine the exact amount of the surplus housing stock in the country through the official statistics, the figures presented above display a significant surplus, most of which is accumulated in the hands of construction firms. These findings are also verified by the news appearing in the media since early 2018 and the housing sales campaigns launched by construction firms and the government.

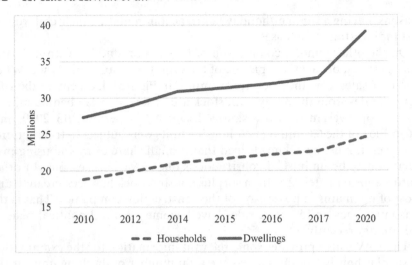

Figure 9.2 The number of dwellings and households in Turkey.

Source: Prepared by the authors based on TURKSTAT (2021c), Kalkınma Bakanlığı (2018), and Strateji ve Bütçe Başkanlığı (2020).

Notes: Figures for the number of dwelling units in 2020 are as of September.

State Intervention: Housing Sales Campaigns and Other Measures

Excess housing production and increasing residential vacancy rates have become a topic frequently covered in the media and a central political issue in the country recently. With the emerging economic crisis in 2018, the government decided to intervene in the housing market and took several steps to prevent a recession in the construction sector. In fact, some of the steps taken in 2018 were not new measures but were applied from time to time to stimulate the economy, such as in 2017. The intervention mainly involved developing tax incentives and cutting mortgage interest rates to increase housing sales. As a first step, in May 2018, the 18% value-added tax (VAT) rate was reduced to 8% on house purchases until the end of October 2018 via a cabinet decree. This tax incentive was extended by Presidential decisions until the end of 2019. Furthermore, in May 2018, the title deed fee was reduced from 20‰ to 15‰ and mortgage interest rates were reduced from 1.25–1.35% to 0.98%.

In addition to these measures implemented by the state, several private real estate organizations in the country came together and announced a housing sales campaign including nearly 100,000 dwelling units in their portfolio. Accordingly, a 20% discount was announced for house purchases from May to June, in addition to a reduction in the down payment from 20% to 5% (Hürriyet Ekonomi, 2018a). As of August 2018, a new housing campaign was launched, this time led by the Ministry of Environment and Urbanization. Similar measures had not been seen previously; the state provided support to (some) private construction firms to enable them to deplete their excess housing stock. This campaign covered 100,000 dwelling units in approximately 400 residential projects (Hürriyet Ekonomi, 2018b). In November 2018, stories started appearing in the media that the amount of excess stock in the country was not known, there were preparations for an inventory, the excess supply was estimated to be around 1 million houses, and the transfer of ready-to-sale houses to the state was on the agenda to solve the deadlock in the housing sector (Habertürk Ekonomi, 2018).

The measures mentioned above are mainly concerned with the problems experienced by the sector's leading companies rather than those experienced by households. These short-term measures developed by the government to stimulate the construction sector will not solve the country's housing problem in the long run. As the discussion above displays, residential vacancy and its dynamics, which are the subject of this paper, are a very significant research and policy area for Turkey. In the light of all these discussions, a case study is presented to evaluate different dimensions and the dynamics of residential vacancy in the Turkish case.

Understanding Residential Vacancy Patterns: A Case Study

Due to the lack of comprehensive official statistics to investigate the residential vacancy issue, a case study was employed to enable further discussion of the residential vacancy dynamics and different types of it in different housing submarkets. For this purpose, the Gemlik District of Bursa Province was selected as a case study area. Field visits to the study area were conducted in 2016 (for more detail, refer to Kıvrak, 2019). Gemlik was selected as the study area due to the practical reason that the Gemlik municipality was cooperative in providing data and information. Moreover, the variety of Gemlik neighbourhoods allows historical housing areas, summer houses, newly developed neighbourhoods, and existing residential areas to be investigated. Furthermore, district-scale is much easier for field visits than city-scale. The information gathered from the field is employed to evaluate the official statistical data for housing stock and households. The purpose here is to discuss the dynamics and different patterns of residential vacancy in detail.

Bursa – Gemlik Case

Bursa is the fourth largest province of Turkey in terms of its population, exceeding 3 million people. In every period of its history, the industrial and agricultural potentials of Bursa have made the city a significant employment centre. Like many other employment centres attracting workforce migration, Bursa is characterized by low levels of authorized housing production compared to the need. Thus, a chronic authorized housing shortage accompanied by high levels of unauthorized housing stock to meet the increasing housing requirement is observed. Bursa's growing economy

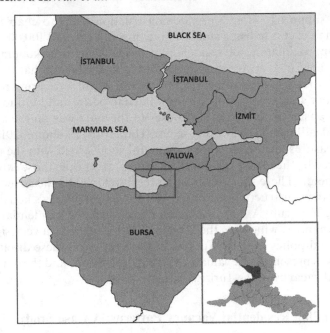

Figure 9.3 Location of Gemlik District and the case study area.

Source: Prepared by the authors based on the raw data provided by the General Directorate of Mapping and Gemlik Municipality.

has also been reflected in Gemlik District. Factories and the Free Zone in Gemlik have a significant role in its development. Furthermore, Gemlik is located by the sea and is in a significant position on the way to İstanbul, the largest employment centre in Turkey (Figure 9.3). Based on its population, Gemlik is the fifth largest district out of 17 and constitutes almost 4% of the population of Bursa Province. There are 35 neighbourhoods in Gemlik, 13 of which are central neighbourhoods that form the case study area of this study. There were almost 43,000 dwelling units and 27,000 households in these 13 neighbourhoods in 2016. The total number of dwellings and households has increased 6% and 11% respectively during the period 2016–2021.

Evaluation of Residential Vacancy Patterns and Dynamics of Vacancy at Neighbourhood Level

In order to estimate the gross vacancy rates at the neighbourhood level, the total number of dwelling units provided in the UAVT are employed and compared with the number of households for the years 2016 and 2021. The number of dwelling units in excess of the number of households is considered the residual. The vacancy rates are calculated as the share of the residual in the total housing stock. Table 9.1 displays the gross vacancy rates

Table 9.1 Neighbourhood level information about vacancy rates, housing stock, and households

Neighbourhoods	Ata	Balıkpazarı	Cumhuriyet	Demirsubaşı	Dr Ziya Kaya	Eşref Dinçer	Halitpaşa	Hamidiye	Hisar	Kayhan	Orhaniye	Osmaniye	Yeni
2016—Gross Vacancy (%)	69.6	30.9	31.4	39.5	23.3	41.2	34.7	33.6	61.3	37.0	38.2	25.1	40.2
2016—Dwellings for sale/rent (%)	1.0	16.6	13.7	10.6	9.7	8.0	6.7	6.9	11.0	8.0	6.2	13.0	8.9
2021—Gross Vacancy (%)	72.4	33.5	30.0	43.3	21.8	32.7	39.8	30.7	49.5	37.9	40.8	26.2	32.9
% change in dwelling units (2016–2021)	5	1	4	0	3	10	3	9	20	3	2	1	1
% change in households (2016–2021)	-4	-3	6	-6	5	26	-5	14	57	1	-2	-1	14
2021—Private workplaces/dwelling units (%)	51	23	2	63	14	12	4	24	11	11	9	10	3
Construction permits issued for residential buildings (2009–2016)	82	4	130	8	256	207	4	275	136	15	2	45	65

Sources: Prepared by the authors based on Kıvrak (2019), raw data from the UAVT, and household numbers obtained from TURKSTAT (2021c).

in 2016 and 2021 based on the UAVT data. As seen, gross vacancy rates range between 23% and 70% in 2016 and between 22% and 72% in 2021. These rates are significantly high compared to the international literature, as expected. As explained above, these high rates are due to the data coverage, which includes all types of dwellings, such as those abandoned and under construction. Furthermore, the data gathered through the field study on the number of dwelling units on offer show that the share of vacant units circulating in the market is lower than the vacancy rates calculated by the UAVT data. However, it must be highlighted that the information about the dwellings offered for sale or rent is just a snapshot of the market reflecting the observation period. Unfortunately, it was not possible to determine the duration of the vacancy since that requires a longer observation period. Thus, it can be argued that the net vacancy rates are somewhere in between these two figures.

Neighbourhood level comparisons from Table 9.1 demonstrate that in some neighbourhoods, for instance, Ata and Hisar, the difference between the overall vacancy and the dwellings offered in the market is considerably high. These two neighbourhoods display similar trends in terms of vacancy; however, the changes in the number of dwellings and households during the period 2016–2021 reveal that they have different narratives. The field study also supports this finding that summer houses dominate the Ata neighbourhood, and Hisar is the development axis of Gemlik and thus has experienced a construction boom recently. The Ata neighbourhood is 5 km from the centre of Gemlik. A Free Zone, industry and port activities, and extensive olive groves occupy most of the area in the neighbourhood. The housing area of the neighbourhood is close to the coast and mainly composed of summer houses. The field study reveals that the housing units in the neighbourhood are not vacant. However, the owners of the summer houses are not registered as part of the neighbourhood population since they reside elsewhere for most of the year.

On the other hand, the Hisar neighbourhood is almost 2 km away from the centre of Gemlik. It was initially a low-density neighbourhood composed of single-family houses and large olive groves. Unlike the neighbourhoods in the centre of Gemlik, Hisar has large areas suitable for new development and has recently attracted construction companies. The neighbourhood was like a construction site in 2016 during the field survey, full of new housing units in five- to seven-storey blocks of flats being prepared to be put on the market. In other words, the vacancy rates in the neighbourhood are due to the new developments. The lower vacancy rates in 2021 and the increasing number of households in the neighbourhood also support this idea (see Table 9.1). However, the exceptionally high vacancy rate in the neighbourhood in 2021 indicates overbuilding. These surplus units are owned predominantly by construction firms.

The second group of neighbourhoods can be identified from Table 9.1 as those with a low (or zero) level of increase in the number of dwelling units accompanied by an adverse change in household numbers during the period 2016–2021. These are the Balıkpazarı, Demirsubaşı, Halitpaşa, Orhaniye, and Osmaniye neighbourhoods. The Kayhan neighbourhood could also be considered in this context, although a minimal increase is observed in household numbers. During the period 2016–2021 vacancy rates in these neighbourhoods have increased. These neighbourhoods are the oldest neighbourhoods in Gemlik and contain historical housing areas as well as archaeological sites. In addition, since 2012, there has been an ongoing discussion about an urban transformation project in Gemlik's centre (affecting these neighbourhoods) due to the earthquake risk associated with the area and the poor quality of construction. Balıkpazarı is a highly dense neighbourhood with poor and medium quality buildings. There are no vacant plots, so demolition of the existing buildings is necessary for new development in Balıkpazarı. The highest rate of for sale or rent dwellings in the area, accompanied by declining household numbers and increased vacancy rates during the period 2016–2021, indicates a low demand for housing in this area. The Demirsubaşı neighbourhood has features similar to those of Balıkpazarı in terms of physical features of the houses, ages of the buildings, and density. The major differences between the two neighbourhoods are that Demirsubaşı has more abandoned buildings and includes significantly more commercial land uses compared to Balıkpazarı. It attracts commercial activities more than residential use. The Halitpaşa neighbourhood, on the other hand, is dominantly a residential area. There are several historical assets in the neighbourhood. The occupied buildings are of low quality and the vacant buildings are abandoned and old. The Kayhan neighbourhood is also an old neighbourhood with limited development capacity. There are historical buildings and archaeological sites within the neighbourhood. Similar to Balıkpazarı, it suffers from low demand.

The Orhaniye and Osmaniye neighbourhoods display features slightly different from those of the other four neighbourhoods. Orhaniye is divided by Kumla Street into two as a coastal settlement and an inland settlement. These two urban parts accommodate different socio-economic groups. The coastal settlement contains good quality buildings, whereas the inland settlement is composed of poor-quality buildings. Several empty houses were observed in the inland settlement during the field visit, not for sale or rent. The demand for housing in this area is low due to the poor quality of buildings and the environment. Osmaniye, on the other hand, is in the vicinity of the cemetery. It includes archaeological sites and historical areas. Although limited, there are new developments in the area. Vacancy in this neighbourhood is observed mainly in the historical parts of the area, where there is low demand. It must be noted that vacant dwellings in the second group of neighbourhoods are owned by individuals, not by construction firms.

The third group of neighbourhoods comprises Cumhuriyet and Dr Ziya Kaya, which have slightly lower vacancy rates during the period 2016–2021 with similar rates of increases in dwelling units and households. These are relatively new and popular neighbourhoods in Gemlik. The Cumhuriyet neighbourhood lies 2 km from the centre. Historically covered by a large area of olive groves, residential construction in the neighbourhood has taken place in the last 20 years. The area has predominantly residential land use, is highly dense and composed of small housing developments, and has a sea view. Therefore, it attracts residential demand and there are plenty of houses for sale and rent in the area. The high vacancy rate observed in the UAVT data is due to the ongoing construction activity in the area. The Dr Ziya Kaya neighbourhood is almost 1.5 km from the centre. It has the lowest levels of vacancy rates in both 2016 and 2021. There is plenty of vacant land in the area and there is a high level of construction activity. As well as residential demand, the area attracts commercial facilities. In the third category of neighbourhoods, both individuals and construction companies own vacant dwellings in differing ratios.

The fourth group of neighbourhoods includes Eşref Dinçer, Hamidiye, and Yeni. All three neighbourhoods are composed of old and new settlements. In the older parts, relatively low demand is observed due to the old and poor quality dwellings. However, a significant amount of construction activity is observed in the new developments in these neighbourhoods. Furthermore, in the Yeni neighbourhood, a considerable number of empty buildings are observed in the older parts of the settlement in 2016. However, a 1% change in the number of dwellings during the period 2016–2021, despite the construction activity in the neighbourhood, implies that some of these old and empty buildings have been demolished in recent years. Like in the third category, vacant dwellings accumulate in the hands of individuals and construction companies in the fourth group of neighbourhoods.

Neighbourhood level evaluations indicate certain patterns in residential vacancy. The first pattern is observed in the older neighbourhoods in Gemlik, where demand for housing is low (e.g., Balıkpazarı). These neighbourhoods are characterized by the low quality of the dwellings, negative expectations of households for the neighbourhood's future, historical housing units that are difficult to maintain, and increasing vacancy rates. This group is an example of both intentional and unintentional vacancy. Buildings that are abandoned (not circulating in the market) reflect intentional vacancy. The owners of these buildings may be under the expectation of speculative profits that could be obtained from the urban transformation in the area. In addition, buildings and residential units that require renovation before they can be occupied are examples of unintentional vacancy. In this case, dwellings are offered for sale or rent, but it could be expected that the duration of the vacancy will be relatively long due to the low demand in the area.

The second pattern is observed in the relatively new neighbourhoods (e.g., Cumhuriyet). The demand for housing is high in these areas. There is an ongoing construction of new dwelling units, yet vacancy rates display a declining trend. This category exemplifies unintentional vacancy, where housing units are habitable, are offered for sale or rent, and the duration

of vacancy will probably be short. The third type is observed in neighbourhoods where old and new stock exists together (e.g., Yeni). In these
neighbourhoods, construction of new dwellings continues either on the
developable vacant land or through demolition and rebuilding. However, in
older parts of these neighbourhoods, intentionally abandoned buildings are
observed as well as buildings that require renovation to make them habitable. The fourth type is seen in the Ata neighbourhood dominated by summer houses, where high vacancy rates observed in the official statistics are
associated with the households not registered in the neighbourhood population. This type is an example of intentional vacancy where the dwellings
are not offered for sale or rent. Furthermore, the final type is observed in the
Hisar neighbourhood, where the supply significantly exceeds the demand,
resulting in overbuilding. Here, although there is demand for housing in the
neighbourhood, due to excess production, the duration of vacancy is long.

Conclusion

This study is an attempt to understand residential vacancy and its dynamics
in Turkish cities. The national literature is limited in this field due to the
lack of data on residential vacancy. This study uses the official statistical
data related to households and housing stock and evaluates them in light
of the insights gained from a case study conducted at the neighbourhood
level. Understanding residential vacancy levels and their dynamics is necessary from the public policy and urban planning perspective. Vacancy rates,
duration of vacancy, and different types of vacancies indicate various problems in the housing market. Therefore, monitoring and managing residential vacancy is a necessity for policy and planning.

The findings of this study indicate that there is a significant surplus of
housing in Turkey. Most of this surplus is accumulated in the hands of construction firms; that is why the government took action to stimulate the
construction sector recently. However, the case study in Gemlik reveals that
vacant dwellings owned by individuals also make up a significant share.
Moreover, the study displays that neither the vacancy rate nor the reason
for the vacancy is uniform across the housing market. In this context, how
to monitor and manage residential vacancy becomes an essential question.

This study provides several clues in the form of inputs for policy and planning. First of all, it is clear that monitoring the changes in the housing stock
is a necessity to manage residential vacancy. Systematic data are required
for tracking residential vacancy in terms of the number and type of vacant
dwellings, duration of the vacancy, and whether these dwellings are fit for
habitation, are offered for sale or rent, and are seasonal housing units or
not. All of this information should be collected at the neighbourhood level.
If such information is not available, it will never be possible to determine
the exact amount of the excess supply in the country and the concentration
of problematic residential vacancy types.

Five different vacancy patterns are identified in the case of Gemlik. Policy and planning responses may prioritize some of them and provide different solutions for them. In low-demand areas where abandoned properties are on the rise due to speculative expectations (Type 1), a vacant property tax could be imposed. This kind of taxation is not implemented currently in Turkey; however, it could be very effective in preventing this type of abandonment. This measure could help planning decisions (such as urban transformation) to be implemented more successfully. If the abandonment is not due to speculation but due to absentee ownership or the need for demolition and rebuilding (Type 3), grants and subsidies could be offered to accelerate renewal. Again in low-demand areas, if the unintentional vacancy is extended for long periods due to the need for renovation (Type 1), to stimulate demand in the area grants or subsidies could be employed, particularly for this kind of property. If the vacancy rates and durations indicate overbuilding in an area (Type 5), then the policy and planning response could be to restrict new development for a certain period in order to maintain the supply–demand balance.

To conclude, monitoring the changes in supply, demand, and their relationship is necessary for the smooth operation of housing markets. Comprehensive data have to be provided to enable research and guide policy. When there is a lack of information, it is not possible to monitor housing problems and develop solutions. Housing problems, like high vacancy levels, are complex and display differences with respect to submarkets. That is why housing-related data at different spatial levels are essential for understanding housing challenges and coping with them.

References

Aksoy Khurami E, Özdemir Sarı ÖB (2020) Özel kiralık kesimde boş konut birimlerinin sosyal konut olarak kullanılması üzerine bir değerlendirme: Cevizlidere mahallesi örneği (Utilization of vacant housing units in social housing provision: An evaluation in the case of Cevizlidere neighbourhood). *İdealkent* 11(3): 1156–1179.

Belsky E (1992) Rental vacancy rates: A policy primer. *Housing Policy Debate* 3(3): 793–813.

Belsky E, Goodman JL (1996) Explaining the vacancy rate-rent paradox of the 1980s. *The Journal of Real Estate Research* 11(3): 309–323.

Couch C, Cocks M (2013) Housing vacancy and the shrinking city: Trends and policies in the UK and the City of Liverpool. *Housing Studies* 28(3): 499–519.

Gabriel S, Nothaft FE (2001) Rental housing markets, the incidence and duration of vacancy, and the natural vacancy rate. *Journal of Urban Economics* 49: 121–149.

Glock B, Häussermann H (2004) New trends in urban development and public policy in eastern Germany: Dealing with the vacant housing problem at the local level. *International Journal of Urban and Regional Research* 28(4): 919–929.

Gu J, Asami Y (2016) Vacant houses, duration for search and optimal vacancy rate in the rental housing market in Tokyo 23 Wards: Based on landlords' optimal search model. *Urban and Regional Planning Review* 3: 31–49.

Habertürk Ekonomi (2018) Konut sektörünü kurtarmanın önemi ve faturası. 07.11.2018, Available at: https://www.haberturk.com/konut-sektorunu-kurtarmanin-onemi-ve-faturasi-2210074-ekonomi#. Accessed 20 May 2021.

Hagen DA, Hansen JL (2010) Rental housing and the natural vacancy rate. *Journal of Real Estate Research* 32(4): 413–433.

Han H (2014) The impact of abandoned properties on nearby property values. *Housing Policy Debate* 24(2): 311–334.

Hillier AE, Culhane DP, Smith TE et al. (2003) Predicting housing abandonment with the Philadelphia Neighborhood Information System. *Journal of Urban Affairs* 25(1): 91–105.

Hoekstra J, Vakili-Zad C (2011) High vacancy rates and rising house prices: The Spanish paradox. *Tijdschrift voor Economische en Sociale Geografie* 102(1): 55–71.

Hürriyet Ekonomi (2018a) Konutta 41 firmanın katıldığı dev kampanya. 15.05.2018. Available at: http://www.hurriyet.com.tr/ekonomi/konutta-dev-kampanya-40837926. Accessed 20 May 2021.

Hürriyet Ekonomi (2018b) 100 bin konutluk dev kampanya. 28.08.2018. Available at: http://www.hurriyet.com.tr/ekonomi/100-bin-konutluk-dev-kampanya-40939854. Accessed 20 May 2021.

Huuhka S (2016) Vacant residential buildings as potential reserves: A geographical and statistical study. *Building Research & Information* 44(8): 816–839.

Jensen JO (2017) *Vacant houses in Denmark: Problems, localization and initiatives.* ENHR Conference 2017 Affordable Housing for All, Tirana, Albania. Available at: https://vbn.aau.dk/ws/portalfiles/portal/262321430/Vacant_Housing_Denmark_ENHR_paper_Jensen_2017_31082017.pdf. Accessed 20 May 2021.

Kalkınma Bakanlığı (2018) *On birinci kalkınma planı (2019–2023): Konut politikaları özel ihtisas komisyonu raporu (Eleventh Development Plan (2019–2023): Housing Policies Specialized Commission Report).* Ankara: Kalkınma Bakanlığı.

Kıvrak HK (2019) *An Investigation of the Residential Vacancy Patterns in an Urban Housing Market: The Case of Gemlik.* Ankara: Middle East Technical University. Available at: http://etd.lib.metu.edu.tr/upload/12624414/index.pdf. Accessed 1 June 2021.

Magnusson TL (2012) Vacancy chains. In: Smith SJ, Elsinga M and O'Mahony LF et al. (eds) *International Encyclopedia of Housing and Home.* Amsterdam: Elsevier, 235–240.

Molloy R (2016) Long-term vacant housing in the United States. *Regional Science and Urban Economics* 59(2016): 118–129.

Özdemir Sarı ÖB (2013) Konut politikalarındaki değişimin mekansal ve toplumsal etkileri: 2002–2013 döneminde Ankara güneybatı koridorundaki gelişmeler (Spatial and social impacts of housing policy shift: Developments in Ankara southwest corridor in the 2002–2013 period). In: *Proceedings of the 4th Symposium of Urban and Regional Research Network (KBAM).* Mersin: Mersin University, 67–80.

Özdemir Sarı ÖB (2015) Konut üretiminde aşırılık ve stok boşluk oranları: Yenimahalle ve Altındağ örnekleri (Oversupply of housing and vacancy rates: The cases of Yenimahalle and Altındağ). In: *Konut (Housing).* Ankara: Şehir Plancıları Odası, 73–84.

Özdemir Sarı ÖB (2019) Redefining the housing challenges in Turkey: An urban planning perspective. In: Özdemir Sarı ÖB, Özdemir SS, Uzun N (eds) *Urban and Regional Planning in Turkey.* Switzerland: Springer, 167–184.

Rosen KT, Smith LB (1983) The price-adjustment process for rental housing and the natural vacancy rate. *The American Economic Review* 73(4): 779–786.

Strateji ve Bütçe Başkanlığı (2020) *2021 yılı Cumhurbaşkanlığı yıllık programı (Presidential Annual Program for 2021)*. Ankara: Strateji ve Bütçe Başkanlığı.

Thalmann P (2012) Housing market equilibrium (almost) without vacancies. *Urban Studies* 49(8): 1643–1658.

TURKSTAT (2021a) Building permit statistics. Available at: https://data.tuik.gov.tr/Kategori/GetKategori?p=Insaat-ve-Konut-116. Accessed 30 May 2021.

TURKSTAT (2021b) House sale statistics. Available at: https://data.tuik.gov.tr/Kategori/GetKategori?p=Insaat-ve-Konut-116. Accessed 30 May 2021.

TURKSTAT (2021c) Address based population registration system results. Available at: https://data.tuik.gov.tr/Kategori/GetKategori?p=Nufus-ve-Demografi-109. Accessed 30 May 2021.

Türel A, Koç H (2015) Housing production under less-regulated market conditions in Turkey. *J Hous and the Built Environ* 30: 53–68.

Vakili-Zad C, Hoekstra J (2011) High dwelling vacancy rate and high prices of housing Malta a Mediterranean phenomenon. *Journal of Housing and the Built Environment* 26: 441–455.

Vandell KD (2012) Housing supply. In: Smith SJ, Elsinga M, O'Mahony LF et al. (eds) *International Encyclopedia of Housing and Home*. Amsterdam: Elsevier, 644–658.

Wood G, Yates J, Reynolds M (2006) Vacancy rates and low-rent housing: A panel data analysis. *J Hous and the Built Environ* 21: 441–458.

10 Discussing Housing Submarkets in the Ankara Metropolitan Area

Tuğba Kütük and Tanyel Özelçi Eceral

Introduction

Housing meets the human need for shelter, being one of the basic human needs, although recent studies have identified other roles of housing, being also a durable consumer good, a source of assurance for the future of families and individuals, an investment tool, and an indicator of status within the contemporary market conditions. These roles assigned to housing have led to the formation of different supply and demand dynamics in the housing market, being unique among the other markets due to its heterogeneity with different spatial, environmental, locational, and physical characteristics. The existence of heterogeneous residential buildings reveals very limited housing groups to be substituted in terms of consumer preferences (Davenport, 2003), and the housing market can thus be divided into many submarkets containing homogeneous groups with similar characteristics.

The housing market plays an important role in modern economies, both at the macro and micro levels. At a macro level, it has a significant impact on the stability of national and international economic systems, considering the contributions of housing construction to production, employment, and investment. The problems in the housing market deepen the effects of economic crises, just as housing markets are negatively affected by economic crises. At a micro level, in addition to meeting the housing need, it constitutes a large part of household spending (Marques, 2012). Changes occurring in the housing market have a considerable impact on the growth and macroform of cities, and the population density in the urban space. Thus, it creates pressure on rural areas and requires the provision of adequate social and technical infrastructure (Marques, 2012). Accordingly, there is a need to make an in-depth investigation of the housing market in order to regulate housing policies and to minimize the mismatch between supply and demand associated with housing policies that focus on housing production. Consumers and producers alike lack in-depth knowledge of the mechanisms behind the housing market, and so the housing market requires government intervention rather than leaving it in the hands of the free market economy. That said, issues related to housing characteristics and the

DOI: 10.4324/9781003173670-13

lack of knowledge of the housing market make the analysis of this issue a complex task (Marques, 2012). Efforts to effectively understand the housing market, the unique characteristics of which make it a complex and spatially diverse system, have led to the emergence of the submarket concept and the idea of determining homogeneous groups.

If the housing market is considered a unitary market, housing policies would be expected to affect a single market, in the same way, ignoring spatial differentiation. Since it is not possible to observe a single housing market, housing policies that do not take into account spatial differences fall short of reflecting the local characteristics of the housing market. The housing market should thus be evaluated as a set of submarkets that include households with similar socio-economic characteristics and with similar spatial and structural dynamics. Approaches that take into account sub-markets contribute to more accurate housing price prediction models, and thus a more precise housing price prediction (Adair et al., 1996; Goodman and Thibodeau, 2007; Wu and Sharma, 2012). There have also been stud-ies suggesting that a submarket analysis can aid in the measurement of the impact of government interventions, which can help governments, city planners, real estate developers, demander groups, and other stakehold-ers make more conscious decisions (Leishman et al., 2013; Keskin and Watkins, 2017).

The present study aims to provide an understanding of the submarkets through an investigation of the case of Ankara, Turkey. In the Turkish context, the construction sector has benefitted from more extensive state support in recent years. In addition to the resulting rapid development of the construction sector, large-scale housing projects developed either by Housing Development Administration (TOKİ) or by the private sec-tor have affected the dynamics of the housing market. The construction of similar types of residential buildings in different cities and the mismatch between housing supply and demand are clear indicators that housing pro-duction has ignored the effects and results of developments in the housing sector. Likewise, individuals' profit expectations from housing investments also increase the demand for housing and affect housing prices. This situa-tion makes any analysis of the housing market in Turkey difficult, requiring an approach that takes into account the effect of supply and demand on housing prices and the conditions and the factors affecting prices. Ankara, as the capital of Turkey, has been one of the cities most affected by the changes in the housing market as a consequence of the housing policies. The mismatch between supply and demand that has resulted from hous-ing policies prioritizing housing production has led to a spatial differen-tiation in the prices of housing with different characteristics in different neighbourhoods in Ankara (Türel, 1981; Alkan Gökler, 2017). Significant differences in house prices can be observed in different parts of Ankara for dwellings with similar structural features. This study aims to show how the house price structure differs in each segment of the housing market, and to answer the question of whether socio-economic submarkets, based on average household income in neighbourhoods, and spatial submarkets, based on the location of the district within the city, are determinants of the Ankara's housing submarkets. To this end, the study investigates whether location, defined by districts, and the average household income by neighbourhood affect the house prices in each submarket or a single market reflects a realistic price structure to understand the housing market structure.

This study is presented in three sections following the introduction. The second section reviews the literature on the definition and classification of housing submarkets and the approaches that determine their dimensions, while in the following section, the data, variables, and applied methodology are described. The fourth section makes a comparison of the empirical performances of alternative submarket models and a market-wide model, after which, some conclusions are drawn from the findings in the last section.

Definition and the Classification of Housing Submarkets

Since the 1930s, the housing submarkets have been an area of considerable interest for urban economists, who have been engaged in debates of its existence and significance since then. The diversity of features of the housing market indicate that it may not be one of the areas of application of the standard price theory (Ellickson et al., 1977). In other words, house prices are not solely tied to the supply-demand relationship. On the other hand, Rosen (1974) argues that the housing market should be evaluated in segments as it does not act as a single market. According to Maclennan et al. (1987), housing stock is divided into separate product groups, with each product group consisting of relatively homogeneous dwellings that represent reasonably close substitutes to the demander group who seek housing. It is how segmented demand is matched with a differentiated housing stock that leads to housing submarkets and results in different prices being paid for certain attributes of properties with similar characteristics in different submarkets (Watkins, 2001).

The concept of substitutability in the housing submarkets is often emphasized by urban economists (Dale-Johnson, 1982; Bourassa et al., 1999; Watkins, 2001; Wu and Sharma, 2012; Pryce, 2013). Substitutability refers to the ability of a house to replace another in the housing sector by providing the same level of benefit to the consumer. If two houses provide the same benefit when used interchangeably, that is, if they can replace each other, it means they are in the same submarket. The existence of this concept in housing can be attributed to the fact that houses are different from other produced goods, having different supply and demand dynamics depending on the buyer and seller profiles and the different price levels (Yiyit, 2017). Patterns of substitutability are linked to the price, location, and structural attributes of the property and the neighbourhood quality (Wu and Sharma, 2012).

The housing submarket concept has been the subject of many studies since Schnare and Struyk's (1976) study of the Boston housing market identifying submarkets based on their inner/outer city location, income, the number of rooms, and the sale and rental market, using the hedonic model developed by Rosen (1974). Studies since then have shown that the residential market can be better analyzed from different but interrelated submarkets set rather than as a single homogeneous market (Watkins, 2001). Bourassa et al. (1999) describe housing submarkets as sets of dwellings that are reasonably close substitutes to one another but relatively poor substitutes for dwellings in other submarkets. Houses that can compete with each other or that have a high rate of substitution, on the other hand, are generally aimed at similar income groups or consumer segments (Grigsby, 1963). Straszheim (1975) suggests that each submarket has different price structures that are based on its specific demand and supply effects, meaning that house prices in a housing submarket reflect the income levels of the households who are willing to pay the going rate for the house on the market. In other words, households with similar economic characteristics choose similar houses due to their similar demands. Furthermore, while lower-income buyers are interested in meeting their basic accommodation needs, higher-income buyers have broader requirements (Greaves, 1984). In this regard, it can be said that analyses of high-income group submarkets are required to be based on different variables to those of low-income groups. In short, a housing market can be defined as a set of submarkets that include homogenous groups with similar socio-economic characteristics and with similar spatial dynamics.

Dividing a large market into submarkets will raise a number of theoretical and methodological questions (Palm, 1978). Despite the consensus on the existence and importance of submarkets, there is no single and coherent definition. Straszheim (1975) and Palm (1978) claim that a submarket is all of the housing in a given geographic area, while Dale-Johnson (1982) suggests that the term refers to all residential units with similar physical properties, regardless of their location, and represent substitutes relatively close to potential buyers. There is also consensus on how submarkets can be determined in practice, although researchers generally agree on a definition based on structural or spatial characteristics (Watkins, 2001). The earliest efforts to distinguish between housing submarkets were based on *a priori* knowledge of census tracts (Straszheim 1975; Ball and Kirwan, 1977; Goodman and Thibodeau, 2003, 2007), postal zones (Goodman and Thibodeau, 2003), school catchment areas (Goodman, 1981; Goodman and Thibodeau, 1998), existing administrative or geographical boundaries (Schnare and Struyk, 1976; Sonstelie and Portney, 1980; Adair et al., 1996; Bourassa et al., 1999; Özus et al., 2007; Uğurlar et al., 2018), and housing submarket transaction data (Dale-Johnson, 1982). The main challenge when using these approaches to identify housing submarkets is the likelihood that the boundaries of submarkets will change over time. In addition to the existing boundaries, the expert knowledge of real estate professionals is also used for defining housing submarkets. (Palm, 1978; Michaels and Smith, 1990; Allen et al., 1995; Watkins, 2001; Keskin and Watkins, 2017).

In recent years, the data-driven approach has come to rely on statistical data analysis techniques, including principal component analyses (Maclennan and Tu, 1996), factor analyses (Dale-Johnson, 1982; Watkins, 1999), cluster analyses (Abraham et al., 1994) or hedonic price models, drawing upon data on multiple housing features simultaneously, and some researchers have applied multiple techniques simultaneously. Bourassa et al. (1999) first used a principal component analysis to determine the factors or variables used to classify submarkets, and then a cluster analysis was conducted to identify homogeneous groups within the large housing market. On the other hand, hedonic pricing techniques are widely used to determine whether price differences exist in the submarkets (Schnare and Struyk, 1976; Goodman, 1981; Dale-Johnson, 1982; Gabriel, 1984; Allen et al., 1995; Adair et al., 1996; Özus et al., 2007; Alkay, 2008). There have to date been many investigating the optimum approach to the modelling of the housing submarkets, although there is still a lack of consensus on the best method for modelling residential submarkets (Rosmera and Mohd Diah, 2016). The data-driven method is more objective than the *a priori* approach, being based on statistical techniques that integrate a wide range of housing feature data. It is also more successful in explaining the temporary dynamics of the housing submarkets, as the data is updated over time. Since the data-based approach cannot spatially limit submarkets, they do not bring spatial proximity to submarkets.

Researchers have identified different dimensions that define housing submarkets. Submarkets are determined by either their spatial or structural characteristics or by both simultaneously. Spatial characteristics refer to the proximity of the housing unit to locational amenities, as well as environmental factors. Some researchers define submarkets in terms of their spatial features, such as their proximity to locational facilities (Goodman and Thibodeau, 1998), the inner and outer urban areas (Schnare and Struyk, 1976; Adair et al., 1996), or the political boundaries (Sonstelie and Portney, 1980). In contrast, there have been researchers who define submarkets in accordance with their structural features, such as the number of rooms (Schnare and Struyk, 1976), floor area (Bajic, 1985), type of housing (Allen et al., 1995; Adair et al., 1996), housing quality (Rothenberg et al., 1991), the number of bathrooms and the age of the building. Other researchers, such as Schnare and Struyk (1976) and Adair et al. (1996), emphasize the combined influence of spatial and structural features in their definitions of housing submarkets. In contrast, there have been studies analyzing the characteristics of demander groups based on such factors as income

(Schnare and Struyk, 1976; Palm, 1978; Alkay, 2008); race and ethnic group (Straszheim, 1975; Palm 1978; Schafer, 1979; Gabriel, 1984); or religious parish (Adair et al., 2000).

The most common approach to the identification of submarkets was introduced by Schnare and Struyk (1976). In this three-step procedure, the first data are first segmented to identify the potential submarkets, after which house price modelling techniques, generally hedonic price modelling, are used to determine the pricing of the different housing submarkets. At this stage, if significant price differences are noted, it is considered evidence of market segmentation. Second, statistical techniques are applied, such as F-tests, to test the significant differences that exist between the specific price estimates of the submarkets. The hedonic price functions of each potential submarket are then estimated to compare housing prices in the potential submarkets. Finally, standard errors are compared to highlight the difference between the hedonic price regression predicted for a single market and the regressions of the potential submarkets. This three-stage test process has been applied either as is or partially modified in subsequent studies (Dale-Johnson, 1982; Gabriel, 1984; Watkins, 2001; Alkay, 2008). A literature review reveals this procedure of Schnare and Struyk (1976) to be the most convenient approach for the purposes of this study but will be modified in line with the study requirements, and the existence of submarkets will be tested. By replicating this process, it will be possible to compare the availability of submarkets in the Ankara housing market.

Evidence from the Ankara Housing Market

Study Area

Ankara, as the capital of Turkey, is one of the county's main cities in which rapid urbanization was experienced with the declaration of the Republic. The rapid population growth brought with it an increasing need for housing, and by the 1930s, Ankara had become a compact city, centred on a single core, and the old stock next to the city centre began to fill up (Alkan, 2015). In the 1950s, the housing stock could no longer meet the housing demand that came with the large-scale migration brought by industrialization. When the provision of housing by the state became inadequate, people began to construct illegal dwellings, known as squatter developments (Uzun, 2005). The enactment of the Flat Ownership Law in 1965 opened up housing production to a build-and-sell model that accelerated the apartmentization process. By the 1970s, high-income groups began choosing locations outside the city, creating satellite settlements after the increase in private car ownership (Uzun, 2005). This led to the sprawl of urban development to the peripheries. During the 1980s, the macroform of the city developed further with the rapidly developing residential areas along the north-south axis, a dense construction pattern, and the massive housing movements that began to appear along the east-west under the guidance of the city master plan. Up until the 1980s, building densities were increased to meet the demand of the growing population, and the 1980s was an important turning point in the promotion of new developments in the planned residential areas and changing housing policies under the influence of neoliberal economic restructuring. In this period, the southwest region became the most speculated and preferred area for the middle-upper and upper-income groups; the northwest region was given over to concentrated massive housing projects that were preferred by the middle and middle-high income groups. The amnesty laws enacted after the 1980s provided for the transformation of the squatter developments, with the aim being to increase urban rents. The applied neoliberal policies, however, differentiated the changing expectations of different social groups with new consumption habits and lifestyles. Since the 1990s, there has been a boom in the construction of ultra-luxurious gated communities by large developers to meet the growing demands of the high-income groups in the major cities of the country (Işık and Pınarcıoğlu, 2009), Ankara included. Entering the 2000s, Ankara has been subject to large-scale spatial changes with the development of the construction sector. Growing numbers of urban transformation and housing renewal projects have taken place, mainly in the older and historical central areas and the former squatter settlements (Ataç, 2016). Housing developments are mostly block-type and high density, and in this period, the city has developed in large pieces with projects concentrated in the southwestern part of the city. After the 2000s, however, it can be stated that the building and population density in the city centre has been increased again with the intensification of transformation projects to deal with the squatter settlements (Alkan Gökler, 2017), concentrated primarily in the eastern, relatively underdeveloped, part of the city, due to the natural boundaries, and preferred mostly by the low-income groups.

Ankara has witnessed different forms of housing supply, such as cooperative housing, mass housing, build-and-sell housing, informal housing, and affordable housing, as the first experimental experience of housing policies in Turkey. With the different residential patterns applied in the different periods, Ankara has become a metropolitan city that is made up of housing submarkets that show similarities or differences in terms of their spatial, economic, and social aspects (Uğurlar et al., 2018). The housing market in Ankara needs to be examined in terms of its submarkets to ensure a better understanding of the current situation. Since the housing market in Ankara has a segmented structure, significant differences in house prices can be seen in stock with the same physical properties and can differ from district to district, but also from neighbourhood to neighbourhood within each district. When examining the housing markets in Ankara, it can be said that the main factor determining the existence of submarkets among different socio-economic groups is price (Alkan, 2015). Thus, a residential segmentation of different income groups can be said to exist in Ankara (Ataç, 2016; Alkan Gökler, 2017). Uğurlar and Özelçi Eceral (2014) mentioned the existence of a housing market in Ankara for the lower- and middle-income groups in neighbourhoods where historical/old residential areas are concentrated in the city, and for the middle-upper and upper-income groups in regions where the central businesses are concentrated. Accordingly, the present study tests the existence of submarkets based on income level and location in Ankara.

The Data and Variables

The dataset of this chapter utilizes three different sources of data. Firstly, the survey data within the Ankara Transportation Master Plan, conducted in 2013, and applied to 43,244 households in eight central districts of Ankara, is employed to calculate the average household income at the neighbourhood level. Secondly, the land value per square metre of the taxable land[1] for 2020 is obtained from the Revenue Administration website.[2] Lastly, non-ready-made data, including the housing, building, and environmental characteristics of the "for sale" housing units, are collected from one of the largest real estate websites in Turkey between September and December 2020.[3]

Among the districts of Ankara, the eight central districts that constitute the metropolitan area, and that are home to the more active parts of the housing market, were selected for the study. For the data collection, a simple random sampling approach was adopted. Studio-type residences, which have only recently gained popularity, and incomplete buildings were excluded from the analysis. The study covers 335 housing units in 160 neighbourhoods of eight central districts of Ankara. The spatial distribution of Ankara's central districts is shown in Figure 10.1.

Data obtained from advertisements for the sale of housing units were evaluated at a district level, providing information on the patterns of substitutability of units based on the indicators of the sale price; the structural characteristics of the dwelling units (floor area, number of rooms, storey on which the unit is located, number of bathrooms and balconies, management fee, heating type, main facade orientation). The data also includes the structural characteristics of the building in question (number of storeys in the building, age, type of dwelling, elevator, garden), some environmental characteristics (part of a housing estate or a stand-alone building, car park, garage, distance to metro, access to public transport, green areas, proximity to primary schools and shopping areas), some locational characteristics of the dwelling unit (land value per square metre, district) and some household characteristics (average household income in the neighbourhood). Each variable is presented in detail in Table 10.1.

Figure 10.1 Location of Ankara's central districts.

Source: Prepared by the authors.

The Method

The existence of submarkets in Ankara in the present study will be tested in two ways, based on the average household income level and the location by district. The relations between the spatial (environmental and locational) and the structural attributes of the dwelling, the building and the household characteristics of the housing market are used as the variables in the analysis. Taking into account the current housing market data, the general characteristics of the central districts of the metropolitan area of Ankara will also be analyzed in the present study, making use of a hedonic model approach to analyze whether the housing price differences in different submarkets are significant. In the hedonic model approach, which is a regression analysis, variables with more than two groups are encoded as dummy variables because they cannot be included directly in the analysis. The variable of management fee has many missing values, and so the arithmetic mean of the existing values of the relevant variable for the missing observations is assigned to this variable for each model. Numerical variables are entered as continuous variables. Prior to the regression analysis, the existence of multi-collinearity between the variables is checked.

If a market is divided into smaller and smaller submarkets, the hedonic prices are less precisely estimated and the standard error will increase (Bourassa et al., 2003). In this direction, Ankara is handled in two stages

Table 10.1 Variable names and definitions

Variables	Coefficients
Price (TRY)	Prices of units for sale between September and December 2020
	HOUSEHOLD CHARACTERISTICS
Income (TRY)	Average household income in the neighbourhood
	LOCATIONAL CHARACTERISTICS OF THE DWELLING
Value (TRY)	Land value per square metre in 2020
Districts	District in which the dwelling unit is located: dummy variables for the Altındağ, Çankaya, Etimesgut, Gölbaşı, Keçiören, Mamak, Sincan, and Yenimahalle districts
	ENVIRONMENTAL CHARACTERISTICS OF THE DWELLING
Site	Dummy variables for environmental characteristics: the
Car park	dummy is 1 if the building is located in a housing estate; if
Garage	there is a car park or parking garage; if in close proximity
Metro	to the metro, public transport, green areas, primary school,
Public transport	and shopping areas
Green areas	
School	
Shopping	**STRUCTURAL CHARACTERISTICS OF THE**
Total storeys	**BUILDING**
Age	Number of storeys in the building
Type	Age of the building in which the unit is located (year)
Apartment	Type of dwelling
	Dummy variables for apartment and single-family house
Single-family houses	Dummy variables for the existence of an elevator, garden
Elevator	
Garden	**STRUCTURAL CHARACTERISTICS OF THE DWELLING**
	Floor area of the dwelling (m²)
Floor area (m²)	Number of rooms in the dwelling
Number of rooms	Storey on which the unit is located
Storey	Number of bathrooms in the dwelling
Bathroom	Number of balconies in the dwelling (terraces are counted as balconies)
Balcony	Monthly maintenance charges
Management fee	The main facade orientation of the dwelling: dummy
(TRY) Facade	variables for north, south, west, and east directions

for comparing potential submarkets model with the same variables. In the first stage, the three main potential submarkets in Ankara's housing market are defined, based on the average household income, making use of an *a priori* classification, with low-income households being those with a monthly income of less than 2500 TRY; middle-income households as those with an income of 2500–5000 TRY; and high-income households as those with an income of over 5000 TRY. The average household income by neighbourhood is preferred, as previous studies have determined that the house prices in a housing submarket reflect the income of the households seeking such properties on the market. In other words, households with similar economic characteristics choose similar houses due to the similar demand dynamics. Likewise, income has a greater impact on the location decisions of households than other housing attributes (Alkay, 2008).

The second stage involves the determination of three spatial potential submarkets based on location by district. House prices vary significantly depending on the location, and housing units in close proximity share similar levels of accessibility to administrative and commercial zones and transport networks, as well as common public services such as schools, libraries, and parks. Although the structural features of the two houses are similar to each other, they cannot be completely the same due to their location (Cingöz, 2014: 93). Türel (1981) determined that house prices in Ankara vary from location to location through the application of a regression model. In his study of rental accommodation, he found that the south of the city was more associated with higher prices and had more differentiated sub-regions than the northern part. Alkan (2015) found that Ankara's residential areas were divided into submarkets in terms of their structural and spatial characteristics so as to meet the demands of households with different socio-economic statuses. Taking the above into account, the existence of submarkets will be tested in terms of their spatial and socio-economic characteristics in the present study, and the two models will be compared with the market-wide model to identify the best strategy for Ankara's housing market in the determination of submarkets.

The spatial distribution of the neighbourhoods included in this study by income groups is presented in Figure 10.2.

In the household survey of 2013 Transportation Master Plan (AUAP, 2014), high-income households could not be determined in the Altındağ, Keçiören and Sincan districts (Table 10.2). More than half of the high-income group reside in the district of Çankaya. In the middle-income group, 40.8% and 25.8% of the sample can be found in the Çankaya and Keçiören districts, respectively. Keçiören, Sincan, and Altındağ contain more than half of the low-income groups. Some of the neighbourhoods in the Gölbaşı district that show rural characteristics are low-income neighbourhoods; on the other side, some of the neighbourhoods located in the prestigious development regions which are close to Çankaya district are high-income neighbourhoods, and there are no middle-income neighbourhoods in the Gölbaşı district.

Secondly, potential spatial housing submarkets are determined by a group of districts according to their location in Ankara metropolitan area, as presented in Figure 10.3. The population of Ankara in 2019 was recorded as 5,639,076. The potential spatial submarket population constitutes 87% of Ankara's total population, with the northeast, the northwest, and the south having shares of 35.4%, 32.1%, and 19.2%, respectively. The northeast region saw the highest population increase in 2019, at 37.2 per mille, compared to the population increases in the northwest region and the south region of 31.6 per mille and 29.4 per mille, respectively.

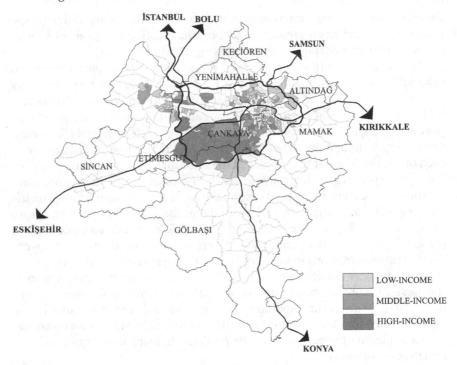

Figure 10.2 Income groups by neighbourhood.

Source: Prepared by the authors based on AUAP (2014).

The potential northeast spatial submarket comprises three districts: Keçiören, Altındağ, and Mamak. The Keçiören district in the north of the city has mostly residential and commercial functions, but also some important service areas, such as university departments and health institutions. The squatter areas in Keçiören were renewed as part of urban transformation projects. The Altındağ district, in the northeast of the city, was the first area to be settled in Ankara, and so contains the historic city centre, while the district of Mamak was one of the main squatter housing areas

Table 10.2 Distribution of average household income by district

Income group (%)	Districts							
	Altındağ	Çankaya	Etimesgut	Gölbaşı	Keçiören	Mamak	Sincan	Yenimahalle
Low income	16.4	7.1	5.0	5.7	23.6	14.3	23.6	4.3
Middle income	8.3	40.8	2.5	0.0	25.8	10.0	9.2	3.3
High income	0.0	64.0	13.3	4.0	0.0	5.3	0.0	13.3

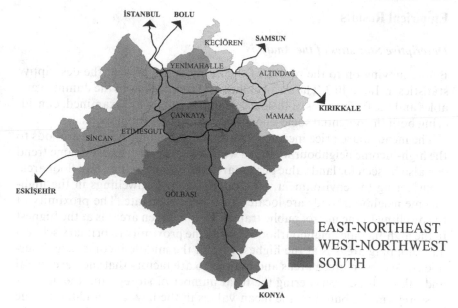

Figure 10.3 Potential spatial housing submarkets by group of districts, according to their location in the Ankara metropolitan area.

Source: Prepared by the authors.

that sprang up during the rapid urbanization of Ankara. Recently, urban transformation projects have spurred activity in the housing sector in the districts of Altındağ and Mamak.

The potential northwest spatial submarket comprises three districts, of which Yenimahalle hosts both developing and established housing areas. Etimesgut, on the other hand, saw much squatter development in the past with the rapid rural-to-urban migration, although the whole area today is planned and has become a development zone. Finally, the Sincan district has based on land-use decisions, resulting in an organized industrial district and a squatter prevention area. The district is today subject to planning implementations related to low-income households.

The potential south spatial submarket south comprises two districts. The first, Çankaya, is home to the main public decision-making bodies, universities, embassies, hotels, cultural and art institutions and is also the most desired residential area. Thus, some of the central business district activities have been spread from the city centre during the last two decades. The second, Gölbaşı, is located on the southern periphery of the metropolitan area and is mostly rural in character, although more recently, residential development has included low-rise housing units for high-income households, as well as high-rise apartments.

Empirical Results

Descriptive Statistics of the Analysis

Before moving on to the results of the analysis, we present the descriptive statistics in Table 10.3. The mean values and frequencies of the dummy variables and the frequencies of the measurable variables are examined, considering both the potential socio-economic and spatial submarkets.

The mean house price increases from the low-income neighbourhoods to the high-income neighbourhoods, as would be expected, and the same trend can also be seen for land value per square metre and dwelling unit floor area. Considering the environment, 60% of the sampled dwellings in the high-income neighbourhoods are located in a housing estate. The proximity of the dwelling to the metro, public transport, and green areas is at the highest in the high-income neighbourhoods, while the proximity to primary schools and shopping areas is at the highest rate in the middle-income neighbourhoods. Access to green areas and the metro are factors that increase rental and sales values. Considering the total number of storeys and the number of storeys in the building, the mean values in the low- and middle-income neighbourhoods are similar but increase in high-income neighbourhoods. Considering the building age variable, it can be said as there is a new housing stock in the low-income group. This can be attributed to the fact that the districts in the low-income group, especially Altındağ and Mamak, are where the urban transformation projects were most diligently applied to alleviate the squatter problem. The fact that the Keçiören district has a more stable housing structure increases the average age of the buildings in the low-income group. Elevators and gardens increase in prominence from the low-income to high-income neighbourhoods, and the floor area of the houses and the number of rooms increases from the low-income to high-income neighbourhoods. Furthermore, it was found that 64% of the houses in the high-income neighbourhoods have a double bathroom.

The mean house price is notably higher in the south than in the northwest and northeast regions, and the mean average household income values follow the same order—south, northwest, and northeast. The land value per square metre is also noticeably higher in the south region. The mean land value in the Northeast region is higher than that of the northwest region due to its advantageous location in terms of accessibility and its proximity to the central business district. The proportion of houses on estates is higher in the northwest region, followed by the southern and northeastern regions. The proportion of single-family houses is also higher in the northwest and south than in the northeast. While the proportion of houses with open car parking is nearly the same for the three potential submarkets, parking garages are more common in the south, as would be expected. The proportion of houses with access to the metro is higher in the northwest region than in the south and northeast regions. Proximity to green areas is highest in the

Table 10.3 Descriptive statistics of potential submarkets and market-wide housing market of Ankara

		Potential socio-economic submarkets			Potential spatial submarkets			Housing market
		Low	Middle	High	North-west	South	North-east	Market-wide
Mean value	Price (TRY)	180,521	350,063	825,293	324,857	613,686	221,613	385,601
	Monthly income (TRY)	1572	3515	6470	3060	4735	2344	3365
	Land value (m²/TRY)	82	244	390	72	422	106	209
	Floor area (m²)	112	126	178	126	150	119	132
	Management fee (TRY)	39	87	286	117	200	37	119
	Storey located	2.5	2.6	5.1	3.70	3.45	2.4	3.1
Existence (%)	Site	14.3	13.3	60	38.1	26.3	13.5	24.2
	Car parking	61.4	38.3	86.7	66.7	66.1	68.4	67.2
	Parking garage	13.6	23.3	50.7	8.9	48.2	20.3	25.4
	Metro	27.1	16.7	40	51.1	22	10.5	26.3
	Public transport	86.4	88.3	94.7	89.3	94.9	83.5	89
	Green areas	72.1	80	85.3	90.5	80.5	67.7	77.9
	Primary school	73.6	76.7	68	89.3	66.1	69.9	73.4
	Shopping areas	54.3	65.8	65.3	72.6	60.2	54.1	60.9
	Elevator	28.6	40	74.7	32.1	56.8	32.6	43
	Garden	26.4	27.5	61.3	41.7	40.7	24.8	34.6
	Single-family houses	0	0	12	6	5.1	0.8	3.6

(*Continued*)

Table 10.3 Descriptive statistics of potential submarkets and market-wide housing market of Ankara (Continued)

Distribution of subcategories (%)		Potential socio-economic submarkets			Potential spatial submarkets			Housing market
		Low	Middle	High	North-west	South	North-east	Market-wide
Total number of storeys in building	1	0	0	0	0	0	0	0
	2–5	90	85	34.7	66.6	69.5	87.5	75.9
	6–9	4.3	10.1	20	16.7	10.2	5.8	19.9
	10+	5.7	4.9	45.3	17.8	20.3	7.7	14.4
Age of building	0–10	33.6	25.8	56	33.4	31.5	41.7	35.9
	11–20	45	40.8	18.7	41.8	36.7	36.2	37.8
	21–30	15.7	22.5	20	16.1	21.9	16.8	18.6
	31+	5.7	10.8	5.3	6	9.9	6.1	7.5
Number of rooms	1	0.7	0	0	0	0.8	0	0.3
	2	32.9	24.2	2.7	22.6	18.7	27.1	23
	3	62.9	70.8	46.7	61.9	55.2	68.4	62.1
	4+	3.6	4.9	50.7	15.5	25.3	4.5	14.6
Number of bathrooms	1	94.3	87.5	25.3	76.2	61.0	90.2	76.4
	2	5.7	11.7	64	21.4	33.1	9.8	20.9
	3	0	0	10.7	2.4	5.1	0	2.4
	4	0	0.8	0	0	0.8	0	0.3
Number of balconies	0	13.6	5	1.3	7.1	8.5	7.5	7.8
	1	70	76.7	60	61.9	71.2	74.4	70.1
	2	13.6	14.2	37.3	28.6	17.8	14.3	19.1
	3+	2.8	5	1.3	2.4	2.5	3.8	3

northwest region, followed by the south region, and there is a considerably low rate in the northeast region, which indicates a high housing density. The proximity to primary school and shopping facilities is also highest in the northwest. While the proportion of six- to nine-storey buildings is greatest in the northwest, followed by the south and northeast, respectively, the proportion of ten plus–storey buildings is higher in the south and northwest than in the northeast. Coming to the building age variable, the proportion of houses in buildings that are fewer than 10 years old is higher in the northeast, followed by the northwest and south, while the proportion of houses older than 20 years is higher in the southern region, followed respectively by the northeast and northwest regions. While elevators are more common in the southern region than in northeast and northwest regions, the proportion of dwellings with gardens is higher in the south and northwest regions than in the northeast region. The mean of floor area is higher in the south than northwest than in the northeast, and in the same vein, the proportion of houses with more than four rooms is also higher in the south, followed respectively by the northwest and northeast. The proportion of houses with more than one bathroom is also highest in the south, followed respectively by the northwest and northeast.

The Market-Wide Model

Previous studies of housing submarkets follow a number of different approaches. House price is a particularly useful variable, being both measurable and a vector of such as the structural, locational, and environmental characteristics of the dwellings. The distribution of housing prices reflects the conditions that are specific to different submarkets, and accordingly, hedonic pricing techniques are used to identify any price differences in the submarkets and the most distinctive features among dwellings. For this reason, the approach is widely used in submarket analysis literature (Schnare and Struyk, 1976; Goodman, 1981; Dale-Johnson, 1982; Gabriel, 1984; Allen et al., 1995; Adair et al., 1996; Bourassa et al., 2003; Özus et al., 2007; Alkay, 2008).

Many regressions were carried out to test different models for Ankara, and the variables found not to be statistically significant are excluded from Tables 10.4 and 10.5. The hedonic model is applied to the entire market-wide housing market, using a linear form, and the results of the regression are presented in Table 10.4. A total of 14 statistically significant variables were found to contribute to the explanation of the house prices in the model. Looking at the standardized coefficient values, these variables can be listed as income (0.51), and the selection of a location in Mamak, Sincan, Keçiören or Altındağ (−0.23, −0.17, −0.16, −0.13), followed by floor area (0.12) and type (0.10). While being in Çankaya and Gölbaşı, where the high-income households are in the majority, don't affect the housing price, other districts have a negative effect on housing price. When considering Ankara as a single

Table 10.4 Hedonic price estimates for market-wide and socio-economic submarket models

| | Housing market | | | Socio-economic submarket models | | | | | | | | |
| | Market-wide | | | High-income | | | Middle-income | | | Low-income | | |
Variables	Unstandardized coefficients B	Standardized coefficients Beta	t	Unstandardized coefficients B	Standardized coefficients Beta	t	Unstandardized coefficients B	Standardized coefficients Beta	t	Unstandardized coefficients B	Standardized coefficients Beta	t
Constant	3.114		29.808	4.034	.185*	5.607	3.778		.553	4.424		10.243
Income	.597	.510***	17.900	.405		2.166	.364	.138*	2.178	.161	.072*	2.226
Altındağ	-.124	-.126***	-5.569				-.083	-.145**	-3.023			
Çankaya										.197	.348***	5.009
Etimesgut	-.117	-.094***	-4.320							.162	.258***	4.538
Gölbaşı										.054	.156**	2.601
Keçiören	-.118	-.157***	-6.333				-.084	-.233***	-4.411			
Mamak	-.217	-.228***	-9.392	-.174	-.303***	-3.595	-.189	-.358***	-6.176	-.068	-.162*	-2.583
Sincan	-.154	-.177***	-6.183									
Yenimahalle	-.104	-.084**	-4.099				-.203	-.230***	-4.592			
Land value	.052	.084**	2.835	.066	.231**	2.979	.134	.373***	6.455	.001	.141*	2.004
Garage				.081	.313***	3.900						
School										-.041	-.124*	-2.116
Total storey	.007	.095***	4.186				.011	.156**	3.100			
Type	.163	.103***	4.607	.104	.261**	2.887						
Elevator	.044	.073***	3.374									
Floor area	.001	.119***	3.290	.001	.297***	3.387	.001	.255***	5.366	.003	.491***	7.956
Rooms	.026	.066*	2.114									
Balcony	.040	.085***	4.473							.038	.179**	2.920

Market-wide: Sample size: 335; R^2: .894; Adjusted R^2: .889; Standard error: .09824; F-value: 192.411

High-income: Sample size: 75; R^2: .635; Adjusted R^2: .603; Standard error: .08200; F-value: 19.721

Middle-income: Sample size: 120; R^2: .761; Adjusted R^2: .744; Standard error: .08062; F-value: 44.257

Low-income: Sample size: 140; R^2: .606; Adjusted R^2: .579; Standard error: .09508; F-value: 22.222

Notes:

$* p < 0.05$ $** p < 0.01$ $*** p < 0.001$

market, on the other hand, this situation supports the presence of higher housing prices in the Çankaya and Gölbaşı districts. The coefficient calculated for each regression indicates the share of the unit amount in housing sales prices, represented by the house attribute variable. Looking at this model, it can be concluded that the independent variables entering the model explain 88.9% (adjusted R^2) of the change in the dependent variable.

Socio-Economic Submarkets

Hedonic prices are estimated for the three potential submarkets, defined according to the average household income by neighbourhood, as the second step. According to the regression results in high-income neighbourhoods, house prices are explained by household characteristics, locational, environmental characteristics, structural characteristics of the building, and structural characteristics of the dwelling. Looking at the standardized coefficient values, it can be seen that the presence of a parking garage has the greatest impact on house prices (0.31), followed by location in Mamak (−0.3), floor area (0.3), house type (0.26), land value (0.23), and income (0.19). Due to the prevalence of single-family houses in high-income neighbourhoods, the housing type in this model can be considered significant.

According to the regression results for the middle-income neighbourhoods, house prices are explained by the household characteristics (income), location (Altındağ, Keçiören, Mamak, Yenimahalle, land value per square metre), the structural characteristics of the building (total storeys), and the structural characteristics of the dwelling (floor area). No environmental characteristics have an impact on housing prices. Furthermore, being located in Mamak, Yenimahalle, Keçiören, or Altındağ has a negative effect on housing price. While land value has the greatest impact on house prices, with a beta of 0.37, income has the least impact, with a beta of 0.14.

According to the regression results in low-income neighbourhoods, house prices are explained by household characteristics (income), location (Çankaya, Gölbaşı, Keçiören, Mamak, land value per square metre), environmental characteristics (primary school) and structural characteristics of the dwelling (floor area, balcony). The structural characteristics of the building have no impact on housing prices in this model, while floor area has the greatest effect on housing prices, with a beta of 0.49, followed by location in Çankaya and Gölbaşı (0.35, 0.26). The differentiations between variables are mostly observable in this submarket.

According to the process devised by Schnare and Struyk, after estimating the hedonic equations, the F-test results are compared to observe whether the implicit prices produced by the hedonic equations of the submarkets for housing attributes are equal or if significant price differences exist (Alkay, 2008). The F-test was applied to three different models and revealed that the coefficients calculated for the different models were significantly different from each other. According to the regression results, the explanatory power of the socio-economic submarket models (adjusted R^2) were respectively 60.3, 74.4, and 57.9% in the high-income, middle-income, and low-income neighbourhoods. The best result was observed in the middle-income model, although when compared to the market-wide model (88.9%), the explanatory powers were lower. In comparison with the market-wide model, the standard errors of the three submarkets were lower in the low-income submarket (0.09508), the high-income submarket (0.08200), and the middle-income submarket (0.08062). Socio-economic models are not linked to all patterns of substitutability. Income, land value, and floor area have a positive effect on housing prices for all models. Mamak emerges as the location of choice as prices decrease for each model and for a single market. While environmental characteristics are insignificant in the market-wide model and middle-income model, they have an impact on housing prices in low- and high-income models. Unlike the other three socio-economic models, being located in Etimesgut and Sincan is significant for a single market and affects house prices negatively.

From an interpretation of the regression results in Figure 10.3, some spatial remarks can be made. While the low-income households are concentrated in the eastern-northeastern part of the city, middle- and high-income households take place more in the southern parts of the city. No direct transition from low-income to high-income can be ascertained in the city in general, aside from in part of the Gölbaşı district and the region within the motorway close to the border of the Çankaya district. This is consistent with the findings of Ataç's (2016) study, in which no status groups showed an even distribution across the city, with the middle-income households acting as a buffer between the low- and high-income households.

The test results presented in Table 10.4 prove that the housing price structure differs in each model. There are three important points that can be deducted from the empirical results. First of all, a single market area price structure does not completely reflect a realistic housing price structure effectively in Ankara, which is consistent with the findings of Alkay (2008) related to the housing submarkets in İstanbul. Secondly, a housing market segmentation based on average household income by neighbourhood can be considered appropriate for Ankara and effective for the Ankara housing market due to its impact on all hedonic equations. On the other hand, groups with similar socio-economic profiles have similar demands and choose similar houses, and so can be evaluated within the same submarkets, according to literature. The present study supports this finding and suggests that submarkets reflect household socio-economic characteristics. Finally, price structure is dependent on different variables in each model. According to Greaves (1984), the determinants of housing prices are parallel to household income that while lower-income buyers are interested in meeting their basic accommodation needs, higher-income buyers have a broader need. In this context, it can be said that the submarkets related to high-income households can be analyzed based on a larger number and a more diverse set of variables than low-income household submarkets, and so it can be expected that the determining coefficient of the submarkets that include high-income household is higher. In the present study, the low-income submarket model has the largest number and the most diverse variables determining the housing prices, contrary to literature. It is thus possible to say that different variables concerning the low socio-economic household have an influential role in the housing market in Ankara.

Spatial Submarkets

Hedonic prices are also estimated for three potential spatial submarkets, defined based on the location of the district, and the regression results of which are shown in Table 10.5. While the west-northwest and south models are linked to all patterns of substitutability, the east-northeast model is linked to household characteristics, location, and the environmental and structural characteristics of the dwelling unit. This approach is more successful than socio-economic submarkets in terms of its linkage of more patterns.

According to the regression results for the west-northwest submarket, income has the greatest impact on house prices (0.67), followed by floor area (0.22), type (0.13), garage (0.1), and land value (0.07). An additional difference can be found in the explanatory power (adjusted R^2) of the west-northwest submarket when compared to socio-economic submarkets, which is higher than the market-wide model. These results can be linked to the low level of distinction between the districts and the more homogenous housing market structure in the west-northwest part. The households in Sincan in particular share similar income levels, and the housing stock has similar structural attributes. For this reason, there are fewer variables that will bring about a differentiation in house prices. On the other hand, housing type has meaningful results because of single-family houses in Yenimahalle.

Looking at the standardized coefficient values, it can be seen that income has the greatest impact on house prices (0.52), followed by type and bathroom (0.18), management fee (0.17), land value (0.13), garage (0.12), site (0.11) and public transport (0.04) in the south submarket. The fact that floor area is not significant in the south submarket model, unlike in other models, suggests that higher prices are paid for houses with small floor areas. In the south part, the proportion of properties on estates is significant, in that households living in the south perceive living in a house located on an estate to be an indicator of status and also a situation that increases the quality of life due to the services offered (such as security, garden, maintenance, etc.). The management fee can also be interpreted as being influential in housing prices due to the existence of many estates in this area. It is interesting that proximity to public transport is meaningful for this submarket, which is home to the majority of high-income households. The existence of low-income households in Gölbaşı and the distance to the city centre could be considered explanations for this result.

Table 10.5 Hedonic price estimates for spatial submarket models

| | Spatial submarket models | | | | | | | | |
| | West-northwest | | | South | | | East-northeast | | |
Variables	Unstandardized coefficients B	Standardized coefficients Beta	t	Unstandardized coefficients B	Standardized coefficients Beta	t	Unstandardized coefficients B	Standardized coefficients Beta	t
Constant	2.230		12.439	3.264		21.730	3.094		21.125
Income	.826	.672***	13.056	.533	.522***	11.610	.417	.462***	9.134
Land value	.062	.069**	2.754	.072	.134***	3.553	.073	.129**	2.650
Site				.053	.105*	2.341			
Garage	.106	.099*	2.448	.053	.117**	2.859	-.043	-.101*	-2.284
Metro							.074	.133**	2.990
Public transport				.038	.038*	2.028			
Green areas							.038	.103*	2.248
Type	.162	.129***	3.345	.179	.176***	4.536	.003	.346***	6.597
Floor area	.002	.224***	4.541				.061	.105*	2.200
Bathroom				.062	.176***	3.744	.040	.152**	3.094
Balcony							.161	.163***	3.672
Management fee				.087	.167***	3.804			

Sample size: 84
R²: .899
Adjusted R²: .893
Standard error: .09797
F-value: 138.821

Sample size: 118
R²: .866
Adjusted R²: .856
Standard error: .08534
F-value: 88.055

Sample size: 133
R²: .766
Adjusted R²: .748
Standard error: .08614
F-value: 44.648

Notes:

* $p \leq 0.05$, ** $p \leq 0.01$, *** $p \leq 0.001$.

House prices are determined by income (0.46), floor area (0.35), management fee (0.16), balcony (0.13), land value and metro access (0.13), number of bathrooms (0.11), green area (0.1), and garage (−0.1) in the east-northeast submarket. The structural characteristics of the building have no meaningful effect on this submarket. In the northeastern part of the city, urban services, resources, and opportunities are limited (Ataç, 2016). While Keçiören has a more stable housing market structure in terms of housing attributes, the markets in Altındağ and Mamak are more active due to the urban transformation of the squatter settlements. In addition to urban renewal projects that affect the low-income households, the location of more prestigious housing increases property values, such as the residences in Mamak (Uğurlar and Özelçi Eceral, 2014). As can be seen, there are different variables from other models in this submarket due to its low socio-economic structure. Proximity to the metro gains importance due to the low rate of car ownership associated with low-income level. The presence of a balcony and proximity to green areas are significant factors related to the high density associated with a high number of dwellings per building. It is somewhat surprising that the presence of parking garage has a negative effect on housing prices; it may be due to low car ownership, which increases housing spending. The largest number and most diverse variables have been observed in this model.

According to the regression results, the explanatory power of the socio-economic submarket models (adjusted R^2) are, respectively, 89.3, 85.6, and 74.8% in the west-northwest, south, and east-northeast areas. In comparison with a market-wide model, the standard errors of the three submarkets are lower in the west-northwest submarket (0.09797), south submarket (0.08534) and east-northeast submarket (0.08614). The best results are observed for the south submarket. The presence of an elevator and the number of rooms are significant in the market-wide model, different from the socio-economic and spatial submarkets. An analysis of the standardized coefficient values reveals income to have the highest effect on housing prices in the three spatial submarkets, which supports the hypothesis that income is an effective factor in the Ankara housing market. The other variable that is significant for the three submarkets is land value per square metre, with house prices being higher in the regions with higher land values. This supports the suggestion that location adds value to housing and leads to an increase in housing prices, consistent with Alkan Gökler's (2017) study. In the submarkets of Yenimahalle, Çankaya, and Gölbaşı that include single-family houses, house type has significant value.

The results of the present study support the suggestion that a spatial differentiation in house prices exists in Ankara based on location (Türel, 1981), meaning that location by districts can be accepted as a determining factor in Ankara for determining submarkets. Spatial submarkets have a higher determinant coefficient than socio-economic submarkets. The fact that the east-northeast submarket has the most variables, where low-income households are in the majority, is different from literature, in a way supports the existence of socio-economic submarkets. These results assert the discrepancy of the Ankara case in terms of the spatial and socio-economic submarkets compared to the various cases reported in related literature.

Conclusion

In the present study, Ankara's housing market is evaluated in terms of patterns of substitutability in the current market value of housing units, the household characteristics (average household income in a neighbourhood), the location (land value per square metre, districts), environmental characteristics (such as car parking, proximity to locational amenities), the structural characteristics of the building (storeys, age, type) and the dwelling unit features (such as floor area, number of rooms). The study aims to show how an analysis of the market as a whole, single market fails to reflect a realistic housing price structure, and that the housing price structure differs between different submarkets, for which the price structure is analyzed in three ways: (i) as a single market, (ii) as socio-economic submarkets, based on the average household income by neighbourhood, and (iii) as spatial submarkets based on the location of the districts within the Ankara metropolitan area. In this regard, the sale price of housing and the average household income that shape consumer behaviours and the different prices paid for houses with similar attributes in different submarkets are taken into consideration in this study.

The present study makes three important findings. First, a single market falls short of effectively reflecting a realistic price structure. Although the explanatory power of the market-wide model is high, it fails to capture the local characteristics of the housing market in Ankara. The study revealed that each submarket has its own dynamics and that the housing price structure in each submarket is dependent on different variables. Furthermore, there are significant price differences between potential submarkets, according to the F-test results, and so it is important to examine the housing market at a submarket level if a better understanding is to be gained. Second, the existence of submarkets is tested in two ways, based on the average household income by neighbourhood; and location by districts with the same variables, so as to ascertain the optimum strategy for Ankara. The results of the analysis of the socio-economic submarkets reveal average household income to be appropriate for the Ankara housing market. Furthermore, location is identified as a determining factor in Ankara in the results of the analysis of the spatial submarkets, which are found to have more explanatory power than socio-economic submarkets. The fact that income has the highest beta value in the spatial submarkets shows that the socio-economic characteristics of households are influential in the Ankara housing market. In addition, the results support the suggestion that households with similar income levels exhibit similar characteristics and can be evaluated as part of the same submarket according to literature. This is effective in the determination of the submarkets in Ankara. Third, the determinants of house prices can be expected to increase as income increases due to the broader needs of the higher-income buyers when compared to the lower-income buyers, who are interested in meeting only their basic accommodation needs. However, in the Ankara case, the number of variables determining house prices differs inversely proportional to income, which contradicts the findings of previous studies. This highlights that the Ankara housing market has a somewhat unique structure that needs to be examined in more detail.

The main limitation of this study is that the date of the survey coincided with the Covid-19 outbreak, which can be expected to affect the balance of the housing market. Other studies related to this situation have reported that house prices have increased, while the reported sales are lower than in the previous year. This study can serve as a base for future analyses of the housing market in Ankara. Determining submarkets can aid decision-makers in focusing on the identification of distinct groups and in directing their resources in the decision-making process. The study can also serve as a source of information on how consumers should use their financial assets. By minimizing the mismatch between supply and demand, the study can help investors make more conscious decisions about housing production based on the current need. The analysis of the housing submarkets, on the other hand, can steer urban planning decisions related to the provision of affordable housing for all income groups in society and in creating settlements in which different income groups are more evenly distributed and share common spaces across the city. Future studies in this vein may also address the rental housing market in a normal period, involving a broader data set.

Acknowledgement

This article is prepared as a part of Tuğba Kütük's ongoing Master's Thesis on Housing Submarkets in Ankara under the supervision of Tanyel Özelçi Eceral.

Notes

1 The land value per square meter of taxable land is based on the minimum land values assessed by the Tax Assessment Committees for the establishment of property tax, which are below the real land values.
2 https://www.gib.gov.tr/sites/default/files/fileadmin/user_upload/ArsaArazi/ANKARA.pdf [Date of Access 20.11.2020].
3 http://www.hurriyetemlak.com/

References

Abraham JM, Goetzmann WN and Wachter SM (1994) Homogeneous groupings of metropolitan housing markets. *Journal of Housing Economics* 3(3): 186–206.

Adair AS, Berry JN and McGreal WS (1996) Hedonic modelling, housing submarkets and residential valuation. *Journal of Property Research* 13(1): 67–83.

Adair AS, McGreal WS, Smyth A et al. (2000) House prices and accessibility: the testing of relationships within Belfast area. *Housing Studies* 15: 699–716.

Alkan L (2015) Housing market differentiation: the cases of Yenimahalle and Çankaya in Ankara. *International Journal of Strategic Property Management* 19(1): 13–26.

Alkan Gökler L (2017) Ankara'da konut fiyatları farklılaşmasının hedonik analiz yardımıyla incelenmesi. *Megaron* 12(2): 304–315.

Alkay E (2008) Housing submarkets in İstanbul. *International Real Estate Review* 11(1): 113–127.

Allen MT, Springer TM and Waller NG (1995) Implicit pricing across residential submarkets. *Journal of Real Estate and Financial Economics* 11(2): 137–151.

Ataç E (2016) A divided capital: residential segregation in Ankara. *METU Journal of the Faculty of Architecture* 33(1): 187–205.

AUAP (2014) Ankara metropolitan alanı ve yakın çevresi ulaşım ana planı hanehalkı araştırması sonuçları 2013. Ankara: Ulaşım Ana Planı Proje Ofisi Yayını.

Bajic V (1985) Housing market segmentation and demand for housing attributes: some empirical findings. *AREUEA Journal* 13: 58–75.

Ball M, Kirwan R (1977) Accessibility and supply constraints in the urban housing accessibility and supply constraints in the urban housing market. *Urban Studies* 14: 11–32.

Bourassa SC, Hamelink F, Hoesli M et al. (1999) Defining housing submarkets. *Journal of Housing Economics* 8: 160–183.

Bourassa SC, Hoesli M and Peng VS (2003) Do housing submarkets really matter? *Journal of Housing Economics* 12(1): 12–28.

Cingöz ARAA (2014) Konut Fiyatları Nasıl Belirlenir? İstanbul: Derin Yayıncılık.

Dale-Johnson D (1982) An alternative approach to housing market segmentation using hedonic price data. *Journal of Urban Economics* 11: 311–332.

Davenport JL (2003) The effect of supply and demand factors on the affordability of rental housing. Honor Projects Paper 10, Illinois Wesleyan University. Available at: http://digitalcommons.iwu.edu/econ_honproj/10. Accessed 12 June 2021.

Ellickson B, Fishman B and Morrison PA (1977) Economic analysis of urban housing markets: a new approach. *Rand Corporation*. Available at: https://www.rand.org/content/dam/rand/pubs/reports/2007/R2024.pdf. Accessed 10 June 2021.

Gabriel S (1984) A note on housing market segmentation in an Israeli development town. *Urban Studies* 21: 189–194.

Goodman AC (1981) Housing submarkets within urban areas: definitions and evidence. *Journal of Regional Science* 21(2): 175–185.

Goodman AC, Thibodeau TG (1998) Housing market segmentation. *Journal of Housing Economics* 7(2): 121–143.

Goodman AC, Thibodeau TG (2003) Housing market segmentation and hedonic prediction accuracy. *Journal of Housing Economics* 12: 181–201.

Goodman AC, Thibodeau TG (2007) The spatial proximity of metropolitan area housing submarkets. *Real Estate Economics* 35: 209–232.

Greaves M (1984) The determinants of residential values: the hierarchical and statistical approaches. *Journal of Valuation* 3: 5–23.

Grigsby W (1963) Housing markets and public policy. Philadelphia: University of Pennsylvania Press.

Işık O, Pınarcıoğlu MM (2009) Segregation in İstanbul: patterns and processes. *Tijdshrift voor Economische en Sociale Geografie* 100(4): 469–484.

Keskin B, Watkins CA (2017) Defining spatial housing submarkets: exploring the case for expert delineated boundaries. *Urban Studies* 54(6): 1446–1462.

Leishman C, Costello G and Rowley S et al. (2013) The predictive performance of multilevel models of housing submarkets: a comparative analysis. *Journal of Urban Studies* 50(6): 1201–1220.

Maclennan D, Munro M and Wood G (1987) Housing choice and the structure of urban housing markets in between state and market housing in the post-industrial era. *Scandinavian Housing and Planning Research* 4: 26–52.

Maclennan D, Tu Y (1996) Economic perspectives on the structure of local housing markets. *Housing Studies* 11: 387–406.

Marques J (2012) *The Notion of Space in Urban Housing Markets.* Unpublished doctoral dissertation, University of Aveiro, Department of Social Sciences.

Michaels R, Smith VK (1990) Market segmentation and valuing amenities with hedonic models: the case of hazardous waste sites. *Journal of Urban Economics* 28: 223–242.

Özus E, Dökmeci V and Kıroğlu G et al. (2007) Spatial analysis of residential prices in İstanbul. *European Planning Studies* 15(5): 707–721.

Palm R (1978) Spatial segmentation of the urban housing market. *Economic Geography* 54: 210–221.

Pryce G (2013) Housing submarkets and the lattice of substitution. *Urban Studies* 50(13): 2682–2699.

Rosen S (1974) Hedonic prices and implicit markets: product differentiation in pure competition. *Journal of Political Economy* 82: 34–55.

Rosmera AN, Mohd DLM (2016) Housing market segmentation and the spatially varying house prices. *Journal of the Social Sciences* 11(11): 2712–2719.

Rothenberg J, Galster G and Butler R et al. (1991) The maze of urban housing markets: theory, evidence and policy. Chicago: University of Chicago Press.

Schafer R (1979) Racial discrimination in the Boston housing market. *Journal of Urban Economics* 6: 176–196.

Schnare A, Struyk R (1976) Segmentation in urban housing markets. *Journal of Urban Economics* 3: 146–166.

Sonstelie JC, Portney PR (1980) Gross rents and market values: testing the implications of Tiebout's hypothesis. *Journal of Urban Economics* 7: 102–118.

Straszheim M (1975) An econometric analysis of the urban housing market. Cambridge, MA: National Bureau of Economics Research.

Türel A (1981) Ankara'da konut fiyatlarının mekansal farklılaşması. *METU Journal of the Faculty of Architecture* 7(1): 97–109.

Uğurlar A, Özelçi Eceral T (2014) Ankara'daki mevcut konut (mülk ve kiralık) piyasasına ilişkin bir değerlendirme. *İdealkent* 12: 132–159.

Uğurlar A, Özelçi Eceral T and Gürel Üçer A (2018) Alt konut piyasaları bağlamında hanehalkı ve konut özelliklerinin ilişkisi. *İdealkent* 25(9): 800–833.

Uzun CN (2005) Ankara'da konut alanlarının dönüşümü: kentsel dönüşüm projeleri. In: Şenyapılı T (ed.) *Cumhuriyet'in Ankara'sı: Özcan Altaban'a Armağan.* Ankara: ODTÜ Yayıncılık, 198–215.

Watkins CA (1999) Property valuation and the structure of urban housing markets. *Journal of Property Investment and Finance* 17: 157–175.

Watkins CA (2001) The definition and identification of housing submarkets. *Environment and Planning A: Economy and Space* 33: 2235–2253.

Wu C, Sharma R (2012) Housing submarket classification: the role of spatial contiguity. *Applied Geography* 32(2): 746–756.

Yiyit M (2017) Isparta ilinde konut fiyatını etkileyen faktörlerin hedonik fiyat modeli ile belirlenmesi ve konut sektöründeki alt piyasaların örtük sınıf analizi ile açığa çıkarılması. Unpublished doctoral dissertation, Süleyman Demirel Üniversitesi Sosyal Bilimler Enstitüsü, Isparta.

11 Do Location Choices of Residents Matter? Comparative Analysis of Two Gated Communities in Bursa

Ebru Kamacı Karahan

Introduction

Gated community is a term well-used in popular and academic literature but the concept is still evolving. In this study, a gated community can be defined "... a physical area that is fenced or walled off from its surroundings, either prohibiting or controlling access to these areas by means of gates or barriers (Landman and Badenhorst, 2018: 218)", and "... they are found from the inner cities to the exurbs and from the richest neighbourhoods to the poorest" (Blakely and Snyder, 1997: 2).

There is a range of issues to be considered when examining a gated community, such as social drivers and built form (Roitman et al., 2010); location choice and residential segregation (Manzi and Smith-Bowers, 2005; Walks, 2014; Özmen, 2020); community development (Alkan Bala, 2018; Colquhoun, 2004; Sanchez et al., 2005; Touman, 2005; Lemanski, 2006; Berköz, 2009; Durington, 2009; Roitman, 2010; Roitman et al., 2010; Spocter, 2012; Obeng-Odoom et al., 2014; Walks, 2014); spatial transformation (Low, 2001; Durington, 2009; Akgün and Baycan, 2012; Landman and Badenhorst, 2012; Li et al., 2012; Ballard and Jones, 2014; Nas, 2017; Alkan Bala, 2018; Landman and Badenhorst, 2018; Chicoine and Whitten, 2019); and housing market (Roitman et al., 2010; Landman and Badenhorst, 2012; Spocter, 2012, 2018). While the literature suggests that the gated community concept requires further research, as Spocter (2018: 330) states, "the core themes of gated living such as security, lifestyle, and prestige are universally present".

Apart from the universal counterparts, in the Turkish context, gated communities seem as not only the prestigious housing estates but also the final products of urban transformation projects in the post-earthquake period (here 1999 Istanbul Earthquakes are referred to), and transform both neighbourhoods and cities in many ways. While gated communities have become a widespread model of urban development used in Turkish cities after the 2000s; a little study focus on gated communities within a broader socio-spatial context, from a comparative point of view (see Geniş, 2007; Landman and Badenhorst, 2012, 2018).

DOI: 10.4324/9781003173670-14

The focus of this paper is twofold: firstly, to examine the development of gated communities in Bursa, the fourth metropolitan city after İstanbul, Ankara, and İzmir; and secondly, to understand living behind the walls of gated housing estates those have different location choices and have different development histories. In other words, this study aims to highlight the dynamics of living in a gated community and interpret the findings in the broader socio-spatial context. Here, the goal is, not only to collect statistical data but also to understand the residents' expectations and perceptions. To gain this understanding, both quantitative (macro-level) and exploratory qualitative (micro-level) research methods are adopted.

The study highlighted this inadequately addressed research area by examining two vertical gated communities in the city of Bursa: Biaport Özlüce on the outskirts and Biaport Zafer in the city-centre. Biaport Zafer and Biaport Özlüce are recent examples of gated communities in Bursa. Three main factors led to selecting them as case study areas: (1) they are two of the few vertical gated communities in Bursa that have rarely been discussed in the literature, contrary to popular examples, such as Koru Park and Harmony Towers; (2) they were built by the same developer, namely Özçeliksan Construction Company, with the almost the same architectural scheme; and (3) they differed from each other in terms of development history, choice of location, residents' socio-demographics, and interactions with the wider context of Bursa. All these features made it possible and meaningful to conduct a comparative study.

The paper consists of four sections. In the next section, gated community developments are discussed in the context of Turkey. The Research Design section gives details about data sources and the methods of the study. That section is followed by a macro-level section, where gated communities are discussed in a broader context, and a micro-level section, where gated communities are discussed in the context of housing estates. In the Concluding Remarks section, the aims and findings of the research are summarized and reconsidered.

Background: Gated Communities in the Turkish Context

Between 1983 and 1991, when she was in power, The Motherland Party (ANAP) sowed the seeds of the wider implementation of neoliberal policies in Turkey with the privatization works carried out. The period that started with the Justice and Development Party (AKP) taking power alone in 2002 was a continuation of the implementation of neoliberal policies that had been taking place since the second half of the 1980s (Recepoğlu, 2018). While the construction industry was seen as the engine of economic growth by the AKP, the government had to prepare an appropriate environment where neoliberal policies operated perfectly. Meanwhile, a strategic document, the Urgent Action Plan (UAP) developed by the 58th Government in 2003, aimed to prevent squatter housing and to transform existing squatter

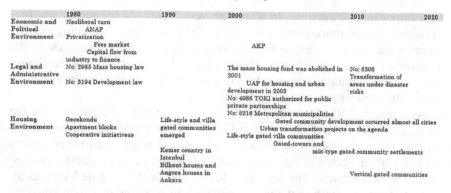

	1980	1990	2000	2010	2020
Economic and Political Environment	Neoliberal turn ANAP Privatization Free market Capital flow from industry to finance		AKP		
Legal and Administrative Environment	No: 2985 Mass housing law No: 3194 Development law		The mass housing fund was abolished in 2001 UAP for housing and urban development in 2003 No: 4986 TOKI authorized for public private partnerships No: 5216 Metropolitan municipalities	No: 6306 Transformation of areas under disaster risks	
Housing Environment	Gecekondu Apartment blocks Cooperative initiativess	Life-style and villa gated communities emerged Kemer country in Istanbul Bilkent houses and Angora houses in Ankara	Gated community development occurred almost all cities Urban transformation projects on the agenda Life-style gated villa communities Gated-towers and	mix-type gated community settlements Vertical gated communities

Figure 11.1 Dynamics of gated community development in Turkey, after 1980.

Source: Prepared by the author.

housing areas into modern urban areas, ensuring planned urbanization by increasing land production, and ensuring that low-income citizens could become homeowners as if they paid rent. Urban transformation projects are presented as the only method to achieve the goals of the UAP (Figure 11.1).

With Law no. 4966 and Law no. 5162, enacted in 2003 and 2004, respectively, the Housing Development Administration (TOKİ) expanded its power over urban development processes and became the most comprehensive and active operator in the housing market. Finally, in 2012, the Law on the Transformation of Areas under Natural Disaster Risk (called the Urban Transformation Law) was enacted, granting TOKİ the right to implement urban transformation projects in the whole country. This accelerated the process of implementing the urban transformation projects whose final products were predominantly gated housing estates. Since the second half of the 2000s, the diverse place- and territory-selective patterns of neoliberal policies on housing markets and urban spheres, such as gated communities, have become a ubiquitous part of urban life.

The construction of Kemer Country and Alkent 2000 in İstanbul, and Bilkent Houses and Angora Houses in Ankara, started the phenomenon of living surrounded by walls in Turkey. Similar to international experiences, this mode of urban development promised security, privacy, and prestige to the residents. As Pérouse (2011) stated, the emergence of new middle or upper-middle classes seeking a lifestyle far from the crowds of the city (Bali, 1999, 2002; Danış and Pérouse, 2005) and the transformation of urban land into a source of rent (Sönmez, 2000) can be counted among the important reasons for the increase in the number of gated communities in Turkey.

Kurtuluş (2011: 52) described gated community development experiences in Turkish cities before 2010 as "the phenomenon of gated communities that indicates new stage in the urbanization experience in Turkey, has become the most popular and attractive housing form for the new urban middle and upper-middle classes". In a similar vein, demands for security, a life-style that can be shared with a similar social group, and the feeling of privilege created by this lifestyle are also seen as reasons for the increase in the number of gated communities. These demands explain the differentiation between lifestyle gated communities in gated villages in the suburbs and vertical gated residences in the city centre.

As Sennett (2018: 2) stated "the forms of the built environment are the product of the makers' [in our case the developers] will". Gated communities are the socio-spatial appearances of a decision-making process driven by the private sector (Cesur Türkmen, 2019). While upper-status groups' preferences determine the location, form, and scope of gated communities, it is safe to state that the transformation of cities was diversified by the types of gated communities. The findings of comprehensive research on gated community development in Turkey conducted by Akgün and Baycan (2012: 95) were that gated communities can be classified into four broad types: "(i) vertical gated developments/gated towers, (ii) horizontal gated developments/gated villa towns, (iii) horizontal gated developments/gated apartment blocks, and (iv) mixed-type gated developments/gated towns". This means that there were not only upper-status American-style gated communities transforming the outskirts of the cities; there were also high-rise residences for the upper status and gated mass housing areas targeting the middle-status groups transforming the city centres (through transforming existing housing areas and developing new housing areas, respectively).

As can be seen in the literature, gated community studies conducted in Turkey mostly choose Istanbul as their case. The main reason for this is that the spatial representation of current economic-political conditions (such as gated communities) are more visible in Istanbul than in the country's other cities. In this regard, Istanbul is followed by the cities of Ankara, Izmir, and Bursa, and this is one of the most important rationales for this study's selecting Bursa as a case.

Research Design

Spatial transformation is informed by particular demands at a specific time (related to the context), and does not happen randomly. While both needs and demands trigger specific ideas about how to address them, form is the physical display of a response to a need and/or demand (Roitman et al., 2010). Such a point of view was a starting point for formulating the mixed-research methodology of the study.

The mixed analysis approach offers a means to interpret gated community developments as part of a much broader socio-spatial transformation process and was formed by building on previous researches, such as that of Batuman and Erkip (2017); Biçki and Özgökçeler (2010); Cesur Türkmen (2019); Ertürk and Karakurt Tosun (2009); Ferreira and Visser (2015); Tanülkü (2013); Tümer and Dostoğlu (2008); Tümer Yıldız and Polat (2011).

To create the main database for macro-level analysis, the work of Recepoğlu (2018), which listed gated community constructed in Bursa until the first half of 2018, was updated by adding gated communities developed between 2018 and August 2020. Here, it is appropriate to restate that, in this study, a gated community is defined as "a physical area that is fenced or walled off from its surroundings, either prohibiting or controlling access to these areas by means of gates or barriers" (Landman and Badenhorst, 2018: 218).

A limited number of studies have focused on the perception and experiences of residents of gated communities and discussed them in a broader context, such as Ibem and Amole (2014); Ibem et al. (2015, 2019); Lemanski (2006); Grant and Mittelsteadt (2004). The discussion has been applied to Turkey by Berköz (2009); Kellekci and Berköz (2006); and Özmen (2020). In order to fill this gap, this study examined the experiences of residents in detail, using an exploratory qualitative research design for micro-level analysis. Data were collected during the fieldwork from two gated communities in Bursa: Biaport Zafer and Biaport Özlüce. During five field visits between February 2019 and July 2020, ten face-to-face semi-structured interviews were conducted. Five people were interviewed in each housing estate and to protect their identities, their names were coded from P1 to P10. Interviewees were asked about their educational profile, profession, reasons for choosing to live in that housing estate, and opinions about the housing estate and the other facilities provided to the residents.

Here, it is necessary to state that, when studying the upper classes, obtaining access in the field is seen as an obstacle (Nader 1972 cited in Tanülkü, 2013). In order to overcome this problem, additional macro-level information was collected from statistics, websites, and previous studies, and a snowball sampling technique was used to identify interviewees for the micro-level analysis. In the following section, the urban development history of the Bursa Metropolitan Area is examined, with special emphasis on gated communities.

Macro-Level: Spatio-Temporal Patterns of Gated Communities in Bursa

Bursa is the fourth most populated city in Turkey, with almost 3.1 million inhabitants (TURKSTAT, 2020). As Recepoğlu (2018) states in addition to its 2200-year history and its characteristics such as being the first capital city of the Ottoman Empire, its geographical location, the combination of agricultural and commercial activities ensure that Bursa always maintains its importance. Besides, the city was a major hub for refugees from various ethnic backgrounds who fled to Anatolia from the Balkans during the Ottoman territories' loss in Europe between the late 19th and early 20th centuries.

Like other industrialized cities in Turkey, Bursa was particularly subject to rapid urbanization and migration until the 1950s. When industrial activities developed in the 1960s, residential areas spread towards the Bursa Plain. Urban development of the city of Bursa is always discussed in relation to the highway to Istanbul in the west, and the Mudanya road and Izmir-Ankara highway in the east. The new settlement areas (firstly squatters and unplanned residential areas) are concentrated around the main transportation axes (motorways) in the northern and western development areas of the city, near the main industrial zones of Bursa (Akın and Erdoğan, 2020).

After the 1990s, a rapid population increase, the growing density in the city centre, and the problems of transportation networks, caused the middle- and upper-class groups to abandon their residences in the city centre (in Altıparmak and Çekirge neighbourhoods) and search for a different habitat. Tümer and Dostoğlu (2008) claimed the residentially mobile upper-class group preferred not to live side by side with the former residents of the city outskirts. So the upper classes leapfrogged the industrial areas that were surrounded by squatters and unplanned housing areas (Akın and Erdoğan, 2020).

The upper-class location preferences and demands for living space in the 1990s were an important turning point that caused the macroform Bursa to take its present form (for further information, see Akın and Erdoğan, 2020). In the 1990s, gated community development started with gated villa settlements located 10–15 km from the central business district (CBD) (Figures 11.2 and 11.3). Bademli Village, located on the old Mudanya road, was the first area in Bursa where upper-class groups settled (Figures 11.3 and 11.4). The village was transformed into a residential area with detached houses, which started with the construction of a site consisting of detached houses (prestigious villas) in 1989 (Tomruk, 2010). As Tümer and Dostoğlu (2008) stated, this transformation occurred after the Development Law No. 3194 (1985) removed the obstacles to the construction of such residential areas.

The newly emerged and liberal cultural, economic, political, and social environment of the 1990s found its spatial counterpart in gated communities. This type of housing production in Bademli Village continued intensively after 2000. The number of gated communities constructed between 1980 and 2000 was 6% of the total number of gated communities constructed by August 2020; the number constructed between 2001 and 2010 was 14%; and 80% were constructed between 2011 and 2020. However, not all gated communities are in the form of villas (Figures 11.2 and 11.3).

As can be seen from Figures 11.2 and 11.5, the development of gated communities tended to be towards the city centre after the 2000s because of the demands of residents. Not only location preferences but also the type of dwelling was diversified to meet their demands (Figures 11.3 and 11.5). By August 2020, the percentage of villas in the total number of gated communities in Bursa was 10%, the percentage of apartments was 82%, the percentage of residences was 5%, and 3% had mixed residential types. Moreover, the location preferences of gated communities were diversified by their residential types. For example, while most of the villa-type gated communities (98% of total) were located on the Mudanya axis, namely Bademli and

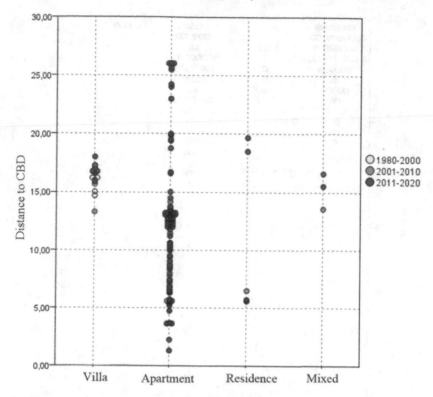

Figure 11.2 Development of gated communities by year, distance to city centre (km) and residential types.

Source: Prepared by the author.

Çağrışan neighbourhoods, 90% of the total residences were located on the İzmir axis, namely Odunluk and Beşevler.

Since the 2000s, the increasing number of these gated housing complexes in Bursa has resulted in a proliferation of urban enclaves, causing uncontrollable urban sprawl, and promoting a lifestyle that is estranged from the city. In order to ensure the regular urbanization of Nilüfer Municipality on the Izmir axis, mass housing projects have been initiated, primarily by the cooperatives (Figure 11.5), and the region has become a place where middle- and upper-class groups prefer to live since the 2010s (Figures 11.3 and 11.4). Between 1980 and 2020, 66% of the total number of gated communities was constructed in Nilüfer District. This means 16 893 housing units were developed in gated communities in this period. The commonest gated community type is apartments, at 90% of the total. Moreover, Nilüfer District hosts 10 out of every 11 prestigious vertical-gated towers (named as 'rezidans' in Turkish) in Bursa. The common features of the gated projects are enclosed secure environments, a common garden, some luxury facilities

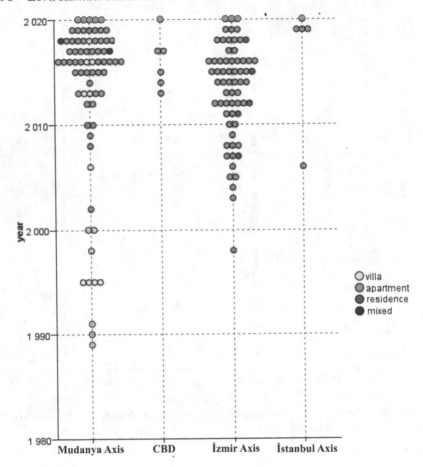

Figure 11.3 Development axis of gated communities by year and type.

Source: Prepared by the author.

(such as an outdoor swimming pool or a closed parking garage), and proximity to motorways.

After the 2010s, the Bademli, Bağlarbaşı, Balat, Çağrışan, Emek, Esentepe, Fethiye, and Yunuseli neighbourhoods on the Mudanya axis also became homes to new gated communities. Between the northern axis of the city and the western axis, there are gated communities from Altınşehir and Ertuğrul on the İzmir road, extending to the districts of Beşevler, Ataevler, Konak, Odunluk, Çamlıca, and İhsaniye. After the latter half of the 2010s, Çalı, Demirci, Özlüce, Görükle, and Kayapa on the Izmir axis, and Hamitler on the Mudanya axis, became preferred locations for gated communities. On the Istanbul road, there are large-scale gated communities in Demirtaş, Millet, and Ovaakça neighbourhoods (Figures 11.3 and 11.4).

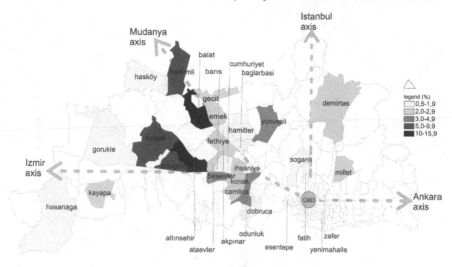

Figure 11.4 Distribution of gated community units by neighbourhoods, 2020.

Source: Prepared by the author.

Gated communities are not evenly spread through the Greater Bursa Metropolitan Area; they tend to agglomerate in the north and north-western parts (splined through the Bursa-Izmir and Bursa-Mudanya axes, respectively) of the metropolitan area (Figures 11.3 and 11.4). By August 2020, the distribution of gated communities through the defined development axes of the city was 50% of the total number of gated communities on the Mudanya axis, 3% in the CBD, 44% on the İzmir axis, and 3% on the İstanbul axis.

The primary results of the macro-level analysis show that gated communities do not directly influence spatial transformation in the entire Bursa Metropolitan Area, but have a significant impact in agglomerated areas. Urban interaction is also likely to vary according to the type of gated community and the proportions of the types in different agglomerations. In Bursa, while large-scale gated communities (such as Biaport Özlüce) tend to threaten the availability of the Bursa Plain's agricultural land on the outskirts of the city (such as Bademli, Kayapa, and Ovaakça), vertical-gated communities (such as Biaport Zafer and Prestij Hayat) in inner-city sites are more likely to accelerate the transformation of low-quality housing areas on the nearby walls into sloppy land.

Micro-Level: Case Studies

Rapoport (1977 cited in Cesur Türkmen, 2019) stated that, to understand a city, one must examine the needs, priorities, demands, expectations, ideals and environments of its residents. This section examines these intangibles as far as possible in the case of two gated communities, namely Biaport Zafer and Biaport Özlüce.

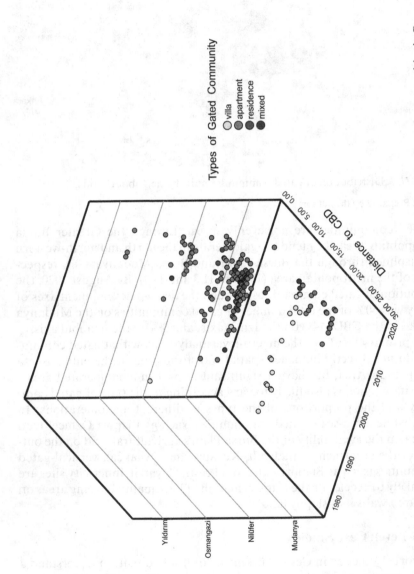

Figure 11.5 Types, construction years, distance to city centre, and districts of gated communities in Bursa.

Source: Prepared by the author.

Case One: Biaport Özlüce

The first project chosen as a case study is Biaport Özlüce located in Özlüce district on the west axis (İzmir Road) of Bursa, close to the Bursa Organized Industrial Zone. Özlüce, the neighbourhood where Biaport Özlüce is located, experienced rapid development through the construction of gated communities ranging from gated villa settlements to apartment complexes (Figure 11.6). The residents of Biaport Özlüce described it as a highly prestigious residence on the road to Bursa Uludag University and Izmir. Those who live in Özlüce are known as secular, well-educated, and leading a white-collar lifestyle. The residents of Biaport Özlüce reflect these characteristics.

As one of the gated housing projects on the periphery of the city, Biaport Özlüce was launched as a "prestigious" housing project for upper-income groups to encourage new lifestyles and increase the comfort of its residents. As the head of the company stated in a news release in 2012: "We [the company] currently have two projects. The total construction area of the project, which will be raised on 20 000 square metres in Özlüce, is 100 000 square metres. We are starting the excavation of our 'Biaport [Özlüce]' project, which consists of 263 flats and 35 shops in 3 blocks; ... There is no taller building other than in Saygınkent. There is no such project in Bursa either. There is a garden, a cafe and a restaurant on the 14th floor. You will be able

Biaport Özlüce

Biaport Zafer

Figure 11.6 Biaport Özlüce and Biaport Zafer.

Source: Google Earth, 2021 and photographed by the author, 2021.

to have your breakfast there in the morning and host your guests there" (Anonymous, 2012).

Biaport Özlüce fulfils the main conditions for being a gated community by offering its residents the security provided by smart home technology, a 24-hour closed-circuit camera system, and private security personnel. It has a playground for children, indoor/outdoor sport areas (gym and pool), green areas, and a sauna. Moreover, there is a reception area, hairdresser, dry cleaner, and cleaning service. The project, when completed in 2014, consisted of 222 flats in 3 blocks with 20 floors. The Biaport Özlüce project includes flats with two, three, and five bedrooms (lofts) with private swimming pools, sky terraces, and social living areas on the 14th floor.

Case Two: Biaport Zafer

In the interview mentioned above, Biaport Zafer Project and the motivation behind the project have been defined as "We will beautify Bursa by demolishing it. And this will happen by urban transformation. ... now we have to protect the central population of the city in Bursa .... we aim to achieve urban transformation with another 'vertical' project of 300 flats that will protect the population of central Bursa" and "Everyone will have to make a suitable project like this project. Thus, Bursa will win. Because that quality will not decrease. ... we bargained with 200 people for two-and-a-half years and bought it for 40 percent less. There are 200 title deeds here". Biaport Zafer was developed by an urban transformation project and is located near the historical city centre. The project was completed in 2017.

Within the scope of the project, two blocks with 180 flats with two, three, and four bedrooms on 15 floors were built on an area of 6730 square metres, which was obtained by demolishing 42 relatively poor-standard buildings. Contrary to the Biaport Özlüce's homogeneous socioeconomic profile, the residents of Biaport Zafer can be defined as heterogeneous. Members of the upper class bought homes from the construction company and the former landowners became homeowners through the urban transformation project, and they all live together in the same housing estate.

Biaport Zafer fulfils the main conditions for being a gated community by offering its residents the security provided by a 24-hour closed-circuit camera system, private security personnel, automatic doors at the garage entrance, a playground for children, outdoor/indoor sport areas (gym and pool), and a sauna. Approximately 2000 people live on this housing estate in flats with two, three, and five bedrooms (duplex).

As can be seen in Figure 11.6, Biaport Zafer is not compatible with the urban texture that surrounds it, in terms of both mass and height. For example, the building completely overshadows the surrounding structures after 14:00 and blocks their sunlight. This voluminosity, seen in the structural sense, is not felt within the site in sociodemographic terms. This could be because it was produced through urban transformation and the former property owners mostly reside on the site. On the other hand, it is relevant to mention here the views expressed during the interviews, such as "we expect this place to be valued" and "we want the surrounding area to be cleaned [referring to the poor-quality building stock]".

Differentiations

In line with Rapoport's (1998) argument that people's different perspectives help us to see details not captured by numbers, these viewpoints provide another platform and context to discuss the urban experience. This section discusses the appropriate data to construct such an approach.

Before proceeding with the comparisons, it would be useful to give brief information about the group interviewed. As can be seen in Table 11.1, the interviewed group comprised ten people. Nine of the ten had university degrees, two of them were academic staff. Regarding the duration of their residency, eight out of ten had lived on their housing estates for more than two years. Among the people interviewed in Biaport Zafer, three of the five were former landlords. In total, four out of ten were tenants, one in Biaport Zafer and four in Biaport Özlüce.

Biaport Zafer is described in the construction firm's promotional website as "Life in the center of the city". This indicates the main objective of the project was to recall people to live in the city centre. But who are the target group? In other words, who are the construction firm calling to the city centre with this project? Most of the interviews in Biaport Zafer claim that the very first responses to this call came from Arabs and Syrians. As P3 and P2 mentioned, on almost all floors of the building, at least one dwelling is occupied by Arabs and/or Syrians. P2—who lives on the eighth floor—puts this situation into words as follows: "Two days before we had

Table 11.1 Profile of groups interviewed

Ref.	Gated community	Age	Education	Job	Former landlord	Owner-ship	Duration of residency
P1	Biaport Zafer	45–50	University	Academic	No	Tenant	2 years
P2	Biaport Zafer	25–30	University	Doctor	No	Owner	1–2 years
P3	Biaport Zafer	30–35	University	Worker	Yes	Owner	2–3 years
P4	Biaport Zafer	45–50	High school	Self-employed	Yes	Owner	2–3 years
P5	Biaport Zafer	65+	High school	Teacher	Yes	Owner	2–3 years
P6	Biaport Özlüce	50–55	University	English teacher	No	Owner	2 years
P7	Biaport Özlüce	65+	University	Academic	No	Tenant	1 year
P8	Biaport Özlüce	45–50	University	Self-employed	No	Owner	3 years
P9	Biaport Özlüce	25–30	University	Engineer	No	Tenant	4 years
P10	Biaport Özlüce	30–35	University	Engineer	No	Tenant	4 years

Source: Prepared by the author.

this meeting with you, three men, possibly Syrian, came out of the next flat, which had been empty for about 3–4 months. There was already a Syrian family on the floor, I said, if I kept in this flat, we would be a minority". On the same topic, P3 whose dwelling is on the second floor, has a differing perception of living with the strangers: "My next-door neighbor is Syrian. Frankly, I don't see any problem living in the same building with my Syrian neighbors. We have good neighborly relations; our wives meet also. They come closer to me than people from my hometown who live on the upper floors".

Here it is the time to ask ourselves, has the call reached the target? One of the interviewers, P4 said: "I was born in a house that was demolished to make this site. We grow up in an Albanian neighborhood as Albanian immigrants. My life here was so safe and blissful. I thought it would be the same; I mean, after we move to our new flats. However, I have not feeling of belonging here. Whenever I see the next-door neighbor, a wealthy Arabic family, I keep asking myself the same question 'why they [refer to the high-status groups of Nilüfer] did not come yet?'"

Biaport Özlüce also has a considerable number of people from European nations and the Middle East. However, the perceptions of interviewees were different from those of Biaport Zafer interviewees. P9, an engineer working for the Bursa branch of an international bank, conveyed his opinion on this subject as follows: "living in Biaport made it possible for me to meet [more] foreign people than I would ever meet anywhere in Bursa. This enriched my personal life as well. If this is not an expression of the prestigious life promised by Biaport Özlüce, what is it?"

Although it was planned that these two estates would have the same social and technical facilities, their current experiences differ from each other. Why did this happen? Variances in the preferences, demands, and lifestyles of residents revealed the reason: Unlike Biaport Özlüce, Biaport Zafer does not have a reception area, open recreational space, or social facilities. P1's feelings about this issue were as follows: "The differences between the two projects can actually be seen on the company's web page. Before we moved to the estate, I googled Biaport Zafer and found the company's promotional web page. In the contrast to Biaport Özlüce Project, Biaport Zafer's web-tab does not serve any information at the page. And after we moved here, I was not surprised to hear that reception does not pass requests to site management. I believe this is one of the consequences of the development model, namely being an urban transformation project".

As mentioned before, when the Urban Transformation Law (no: 6306) was enacted in 2012, construction companies turned their attention to valuable but low-quality inner-city areas. In Bursa, five gated communities were developed by urban transformation projects (Bulvar224 and Begonvil Park Houses on Izmir road; Biaport Zafer in the city centre, Atış Premium in Soğanlı, and Prestij Park in Istanbul axis). Approximately 2156 housing units were constructed between 2016 and 2020. While these projects constitute only 16% of the dwellings constructed in the period (n:13 507 flats), they can be considered as preliminary projects for the restructuring of the city centre. In agreement with this view, people interviewed believed that Biaport Zafer has the power to change the character of the area. This belief was expressed by P4 as follows "our old neighborhood, full of low-quality housing, will become a preferred housing area for the ones who live in Nilüfer today, sooner than expected. The days when we will see this transformation will come soon".

Although the development history of the projects was different, the design of Biaport Zafer reminds one of Biaport Özlüce, viewed from the outside. P2 and P5 stated their opinions respectively as follows: "I went to my brother-in-law, who lived in Biaport Özlüce, and I truly say, because of our housing estate was developed by urban transformation, the floor plans of the dwellings are not well-designed" and "when I first saw my apartment, I was shocked because there was a column in the middle of the dining room. But, after 3 years, we are used to living with it". As a homeowner in Biaport Özlüce, P8 said this: "In fact, we were tenants on this site before we bought this house. We wanted to buy a house here because we like the plans of the houses. And large terraces and long balconies are one of the main reasons why we like this house".

While there is no statistical evidence, the micro-level analysis provided the first meaningful hints that these housing estates are differentiated by the socioeconomic and demographic composition of the residents, their motivations for living there, their perceptions of the estates, and their expectation of living in a gated community. The people living in Biaport Özlüce reflect the same socioeconomic and demographic profile as those in Biaport Zafer. Biaport Özlüce residents have similar priorities and judgments, which increases their sense of belonging. While the main motivation for living in Biaport Özlüce is the desire to live in a safe and prestigious community, the desire to be close to work is a primary reason for residing in Biaport Zafer. This profile was seen in both newcomers to and former landlords of the site. Because the former landlords saw the new flats as a source of rent and they believed the right time had come, they moved to a place where people like themselves lived. However, their unspoken intentions were also reflected in newcomers' sentiments: "They cannot pay the maintenance fee of this site for long. Eventually, they move" (P2).

Concluding Remarks

Gated communities have been seen as accepted development strategies in both developing and developed countries. Today, we are living in a period in which the gated communities' share in the housing market is increasing rapidly, not just in Turkey's metropolitan areas, such as Istanbul, Ankara, and Bursa, but in almost all cities. Both the upper classes and the new middle-upper classes are their target consumers.

Gated communities have been a good subject for researchers since they provide a rich vein for research in terms of urban theory and empirical profiling, along with providing the desired housing in our towns and cities. Especially in developing countries, where the construction industry is one of the key elements of neoliberal development, there is a need to conduct studies on the perceptions of users when choosing a habitat.

As previously stated, this study is about examining the internal dynamics of living in two gated communities with different locations and development histories; and interpreting the findings within the broader scale of the socio-spatial transformation of Bursa. Having been built on the findings of an exploratory qualitative field research, this paper aims to fill this gap partially and to pave the way for further studies on the phenomenon.

The macro-level analysis shows that the residential types, development history, and location preferences of gated communities were differentiated in the period between 1990 and 2020. By the 2010s, gated community development in Bursa had entered a new period: the number of gated communities almost doubled, the location choices diversified, and the scope and volume of the projects varied considerably. Those projects were generally prestigious gated communities that were enclosed, secured, and offered improved lifestyle and social opportunities within their borders. Not only did gated communities begin to serve different parts of society, but also the inner and the outer parts of the city became eligible for investment in the urban development project (model).

As pointed out in the study, living behind walls is not unique to upper-status groups today. Since the Urban Transformation Law was enacted in 2012, urban development strategies and the patterns of most of Turkey's cities have been brought to life by urban development projects. Using the legal, operational and financial means provided by the state, construction firms make gated communities for all segments of society. However, with a diversified development history, the residents' socioeconomic, demographic, and ethnic composition differ, as do their interactions with the urban environment. In this respect, the findings of the macro-level analysis and the explanatory quantitative micro-level analysis both provided clues about the importance of examining not only the type and standards but also the location and development history of gated communities for further research.

References

Akgün AA, Baycan T (2012) Gated communities in Istanbul: The new walls of the city. *Town Planning Review* 83(1): 87–109.

Akın A, Erdoğan MA (2020) Analyzing temporal and spatial urban sprawl change of Bursa city using landscape metrics and remote sensing. *Modeling Earth Systems and Environment* 6(3): 1331–1343.

Alkan Bala H (2018) Neighborhood from cul-de-sac to gated community in Turkish urban culture: The "Fina". *ICONARP International Journal of Architecture and Planning* 6(2): 333–357.

Anonymous (2012). Osman Çelik kentsel dönüşüme talip. *Ekonomi*, 22 May. Available at: https://biaport.com/osman-celik-kentsel-donusume-talip.html. Accessed 20 July 2021.

Bali RN (1999) Çılgın Kalabalıktan Uzakta ... *Birikim Temmuz* 123: 35–46.

Bali RN (2002) *Tarz-ı Hayattan Life Style'a: Yeni Seçkinler, Yeni Mekânlar, Yeni Yaşamlar.* İstanbul: İletişim Yayınları.

Ballard R, Jones GA (2014) The Sugarcane Frontier: Governing the Production of Gated Space in Kwazulu-Natal. In: Haferburg C, Huchzermeyer M (eds.) *Urban Governance in Post-Apartheid Cities: Modes of Engagement in South Africa's Metropoles, Urbanization of the Earth.* Stuttgart: E Schweizerbart and Gebr Borntraeger, 295–312.

Batuman B, Erkip F (2017) Urban design—or lack thereof—as policy: The renewal of Bursa Doğanbey district. *Journal of Housing and The Built Environment* 32(4): 827–842.

Berköz L (2009) Comparing the residential developments in gated and non-gated neighborhoods in Istanbul. *A|Z ITU Journal of Faculty of Architecture* 6(1): 41–59

Biçki D, Özgökçeler S (2010) Gentrification in question: The case of Bursa, Turkey. 2nd International Symposium on Sustainable Development, Sarajova: 436–444.

Blakely E, Snyder M (1997) *Fortress America: Gated Communities in the United States.* Washington DC: Brookings Institution Press.

Cesur Türkmen SÇ (2019) Changing urban environment and lifestyles: A study on housing enclaves in Bursa. Unpublished master thesis. Middle East Technical University.

Chicoine D, Whitten A (2019) 6 Gated Communities, neighborhoods, and modular living at the early horizon urban center of Caylán, Peru. *Archeological Papers of the American Anthropological Association* 30(1): 84–99.

Colquhoun I (2004) Design out crime: Creating safe and sustainable communities. *Crime Prevention and Community Safety* 6: 57–70.

Danış AD, Pérouse J (2005) Zenginliğin mekânda yeni yansımaları: İstanbul'da güvenlikli siteler. *Toplum ve Bilim* 104: 92–123.

Durington M (2009) Suburban fear, media and gated communities in Durban South Africa. *Home Cultures* 6(1): 71–88. DOI: 10.2752/174063109x380026.

Ertürk H, Karakurt Tosun E (2009) Küreselleşme Sürecinde Kentlerde Mekânsal, Sosyal ve Kültürel Değişim: Bursa Örneği. *Uludağ Üniversitesi Fen-Edebiyat Fakültesi Sosyal Bilimler Dergisi* 10(16): 37–53.

Ferreira V, Visser G (2015) A spatial analysis of gating in Bloemfontein, South Africa. *Bulletin of Geography* 28(28): 37–51.

Geniş Ş (2007) Producing elite localities: The rise of gated communities in Istanbul. *Urban Studies* 44(4): 771–798.

Grant J, Mittelsteadt L (2004) Types of gated communities. *Environment and Planning B: Planning and Design* 31(6): 913–930.

Ibem EO, Adeboye AB and Alagbe OA (2015) Similarities and differences in residents' perception of housing adequacy and residential satisfaction. *Journal of Building Performance* 6(1): 1–14.

Ibem EO, Amole D (2014) Satisfaction with life in public housing in Ogun State, Nigeria: A research note. *Journal of Happiness Studies* 15(3): 495–501.

Ibem EO, Ayo-Vaughan EA and Oluwunmi AO et al. (2019) Residential satisfaction among low-income earners in government-subsidized housing estates in Ogun State, Nigeria. *Urban Forum* 30: 75–96.

Kellekci ÖL, Berköz L (2006) Mass housing: User satisfaction in housing and its environment in Istanbul, Turkey. *European Journal of Housing Policy* 6(1): 77–99.

Kurtuluş H (2011) Gated communities as a representation of new upper and middle class in İstanbul. *İ. Ü. Siyasal Bilgiler Fakültesi Dergisi* 44: 49–65.

Landman K, Badenhorst W (2012) The Impact of Gated Communities on Spatial Transformation in the Greater Johannesburg Area. South African Research Chair in Development Planning and Modelling, School of Architecture and Planning, University of the Witwatersrand, South Africa.

Landman K, Badenhorst W (2018) Gated Communities and Spatial Transformation in Greater Johannesburg. In: Harrison P, Got G, Tode A et al. (eds.) *Changing Space, Changing City: Johannesburg after Apartheid.* Johannesburg: Wits University Press, 215–229.

Lemanski C (2006) Spaces of exclusivity or connection? Linkages between a gated community and its poorer neighbor in a Cape Town master plan development. *Urban Studies* 43(2): 397–420.

Li SM, Zhu Y and Li L (2012) Neighborhood type, gatedness, and residential experiences in Chinese cities: A study of Guangzhou. *Urban Geography* 33(2): 237–255.

Low SM (2001) The edge and the center: Gated communities and the discourse of urban fear. *American Anthropologist* 103(1): 45–58.

Manzi T, Smith-Bowers B (2005) Gated communities as club goods: Segregation or social cohesion? *Housing Studies* 20(2): 345–359.

Nader L (1972) Up the Anthropologist: Perspectives from Studying up. In: Hymes D (ed.) *Reinventing Anthropology.* New York: Pantheon Books, 284–311

Nas A (2017) Between the urban and the natural: Green marketing of Istanbul's gated community projects. *İdealkent* 8(22): 396–422.

Obeng-Odoom F, Eltayeb EYA and Jang HS (2014) Life within the wall and implications for those outside it: Gated communities in Malaysia and Ghana. *Journal of Asian and African Studies* 49(5): 544–558.

Özmen EE (2020) The perception and experience of spatial segregation: Dikmen 5th stage gecekondu neighborhood and park Oran gated community. Unpublished master thesis, Middle East Technical University.

Pérouse JF (2011) *İstanbul'la Yüzleşme Denemeleri: Çeperler, Hareketlilik ve Kentsel Bellek.* İstanbul: İletişim Yayınları.

Rapoport A (1977) *Human Aspects of Urban Form: Towards a Man – Environment Approach to Urban Form and Design.* Oxford: Pergamon Press.

Rapoport A (1998) Using "culture" in housing design. *Housing and Society* 25(1–2): 1–20.

Recepoğlu S (2018) Türkiye'de kentsel mekanın dönüşümü: Bursa örneği. Unpublished master thesis, Adnan Menderes University.

Roitman S (2010) Gated communities: Definitions, causes and consequences. *Urban Design and Planning* 163(1): 31–38.

Roitman S, Webster C and Landman K (2010) Methodological frameworks and interdisciplinary research on gated communities. *International Planning Studies* 15(1): 3–23.

Sanchez TW, Lang RE and Dhavale DM (2005) Security versus status? A first look at the census's gated community data. *Journal of Planning Education and Research* 24(3): 281–291.

Sennett R (2018). *Building and Dwelling.* UK: Penguin Books.

Sönmez M (2000) İstanbul'da kuzey güney kutuplaşması ve rantlar. *İstanbul* 35: 105–108.

Spocter M (2012) Gated developments: International experiences and the South African context. *Acta Academica* 44(1): 1–27.

Spocter M (2018) A toponymic investigation of South African gated communities. *South African Geographical Journal* 100(3): 326–348.

Tanülkü B (2013) Gated communities: Ideal packages or processual spaces of conflict? *Housing Studies* 28(7): 937–959.

Tomruk B (2010) Bursa'nın 2000–2010 Arası Yeniden Yapılanmasında Kentsel Söylem Üzerinden Dönüşüm Rotaları. Unpublished doctoral thesis. İstanbul Technical University, Turkey.

Touman AH (2005) Gated communities: Physical construction or social destruction tool? Available at: https://citeseerx.ist.psu.edu/viewdoc/download?doi=10.1.1.582.9503& rep=rep1&type=pdf. Accessed 22 July 2021.

Tümer HÖ, Dostoğlu N (2008) Bursa'da dışa kapalı konut yerleşmelerinin oluşum süreci ve sınıflandırılması. *Uludağ Üniversitesi Mühendislik-Mimarlık Fakültesi Dergisi* 13(2): 53–68.

Tümer Yıldız HÖ, Polat S (2011) Bursa-Korupark alışveriş merkezi ve Korupark evleri'nin mekânsal, anlamsal ve göstergebilimsel analizi. *Uludağ Üniversitesi Mühendislik-Mimarlık Fakültesi Dergisi* 6(2): 11–24.

TURKSTAT (2020) *Address based Population Census Results*. Available at: https:// data.tuik.gov.tr/tr/Display-Bulletin/?Bulletin=Adrese-Dayali-Nufus-Kayit-Sistemi-Sonuclari-2019-33705#. Accessed 22 July 2021.

Walks A (2014) Gated communities, neighborhood selection and segregation: The residential preferences and demographics of gated community residents in Canada. *Town Planning Review* 85(1): 39–66.

12 The Repair, Maintenance, and Restoration of Traditional Housing and the Related Legal Framework

Antalya Kaleiçi

Aynur Uluç

Introduction

Residential areas are formed under the influence of the physical, social, technological and cultural factors in place at the time of their construction. However, lifestyles and needs change in time with modernization and technological developments (Medici et al., 2019; Halifeoğlu et al., 2020), and traditional housing areas are subsequently no longer able to meet the needs of communities (Yetgin, 2005; Saka and Özen, 2013; Mısırlısoy and Günçe, 2016). With the passage of time, interventions into the historic urban fabric become inevitable for conservation and ensuring the protection of the exterior appearance, for the transforming of the interior spaces to meet the requirements of modern life, and for the gradual improvement of the local infrastructure and environment (Wang, 2012: 42–43). Appropriately, conservation is based on facilitating repair and restoration work if necessary, providing grants, and ensuring the integrity and authenticity of heritage (Stovel, 2002: 36–37 cited in Elsorady, 2011: 500). Although adapting traditional housing units to modern living conditions and maintaining them necessitates a certain level of intervention, there are some challenges faced by the current residents of listed traditional housing units in Turkey, including the limitations associated with the legal framework, the lack of finances to maintain, repair and restore their housing units, the difficulties in cases where there is more than one property owner and the long bureaucratic processes associated with interventions.

According to the Law on the Conservation of Cultural and Natural Property (Law No. 2863), immovable property with cultural and historical significance that was built until the 19th century, or those with special characteristics, despite being built after the 19th century, are to be considered "cultural property to be conserved". In Turkey, listed or registered buildings are those registered as historic buildings either as a single cultural property or as part of a designated conservation site, or both, in the National Registry. Law No. 2863 assigns responsibility for the registration of cultural properties to the General Directorate of Cultural Properties and Museums, a division of the Ministry of Culture and Tourism. A listed building obtains

DOI: 10.4324/9781003173670-15

the status of a public good regardless of whether or not it is privately owned, which brings limitations and controls to the interventions residents can make to the property (Carter and Grimwade, 1997; Ulusan and Ersoy, 2018; Jahed et al., 2020). Development rights related to such properties are also prohibited (Jahed et al., 2020), bringing additional burden to the current residents of housing units, resulting in listed historic buildings[1] being considered in a negative light by their owners.

The financial side of interventions into listed historic buildings can also be evaluated as problematic, since the repair,[2] maintenance, or restoration costs may be restrictively high and beyond the means of their users or owners (Madran and Özgönül, 2005: 48). This problem can be seen particularly in listed historic buildings whose owners are in the lower–middle-income and low-income bands, and such buildings are inevitably left to their fate (Ulusan and Ersoy, 2018; Jahed et al., 2020). Disinvestment in such buildings leads to a decline in their physical condition and economic value, resulting over time in urban decay (Medici et al., 2019: 90), and delaying the repair, maintenance, and restoration of historic buildings increases costs over time, as the problems of buildings get worse if they are not fixed promptly (Şahin, 2006). Furthermore, the lack of sufficient resources to repair, maintain and restore their housing units may lead to owners opting to sell them for transformation into commercial and touristic facilities.

Although some direct financial support is available, such as grants from the Ministry of Tourism and Culture, loans from the Housing Development Administration (TOKİ), and grants provided by the governorate sourced from property taxes, the proffered funds seem to fall short of covering the restoration costs. The absence or insufficiency of funding sources has reached such dimensions that in some cases, even the simplest interventions required for the survival of the buildings are neglected (Madran and Özgönül, 2005: 48). An additional challenge relates to there being more than one property owner as a result of a shared inheritance, and identifying all of the owners is difficult in some cases. Restoration processes are subject to the permission of all owners, and so such housing units may often be left to ruin (Uslu, 2018b).

The current legislation governing the repair, maintenance, and restoration of listed historic buildings in Turkey also brings challenges to the residents of such buildings, including the complex bureaucratic process involved. Even the most basic maintenance and repair, restoration, or reconstruction require the permission and supervision of the Regional Conservation Council, as indicated in Law No. 2863, falling under the responsibility of the Conservation, Implementation and Control Office (KUDEB). Residents must apply to these bodies to obtain a Preliminary Permission Certificate for Repair and a Conformity Certificate for Repair for any required interventions. All of the above contribute to a situation in which housing units in a state of disrepair, both vacant and in use, can be observed in many historic urban areas.

The maintenance of the original function of historic structures with their residents has been addressed in different international documents. Among these, the Declaration of Amsterdam emphasizes that the rehabilitation of historic urban areas should not necessitate a major change in the social composition of the residents, and that all sections of society should share the benefits of restorations financed by public funds (CoE, 1975). The Washington Charter, on the other hand, calls for the involvement of residents at the outset of efforts to conserve historic areas (ICOMOS, 1987), while in the Valletta Principles, the deterioration of towns and housing was emphasized (ICOMOS, 2011). As can be understood from all the above, the existence of host communities and their use of historic buildings has become a significant matter (Ecemiş Kılıç and Türkoğlu, 2015; Yung and Chan, 2012), as historic urban areas gain meaning from their authentic use and users. Use also contributes to the conservation of such structures in, for example, the United Kingdom, where vacant listed buildings are considered at risk when left to decay (Historic England, 2011).

Traditional housing stock is also infused with its inhabitants' social habits and cultural values, whose sustainability is also significant for cultural continuity (Mısırlısoy and Günçe, 2016). The sustainability of housing and its residents both contributes to the conservation of the traditional urban pattern, and contributes to the resolution of the problem of housing demand, in that the use of existing housing stock with its established infrastructure is preferable to the construction of new housing stock. The rehabilitation of traditional and other building stock, as was seen in Montreal and France, led to housing being provided in city centres (Arabacıoğlu and Aydemir, 2007: 211). Making use of existing buildings rather than constructing new ones brings benefits to existing locations in urban districts by decreasing construction costs and planning fees (Medici et al., 2019: 90), given the usual proximity of such areas to the city centre. Policies that prevent traditional buildings from being used to meet housing demand, and that focus rather on new construction, fail to contribute to conservation and ignore the high economic and functional value of traditional building stock (Madran and Özgönül, 2005: 57). The success of conservation depends on the protection of the landscape, function, and culture together, preventing possible conflicts in the community, meeting the needs of the residents, decreasing the responsibility of the government, and facilitating cooperation between the government and the public (Wang, 2012: 43). Elsorady (2011) highlights the need to maintain a balance between the needs of the owners and the legal framework for the successful conservation of cultural heritage, and to ensure the necessary implementations and the provision of the required tools for the conservation of cultural heritage. Although her research was developed around Egypt's legal framework, the same applies to the Turkish case.

The present study addresses the challenges faced by the residents of traditional housing units related to repair, maintenance and restoration, and analyzes the related legal framework in Turkey. To this end, an analysis is made of the current housing stock in the Kılınçaslan neighbourhood in Antalya Kaleiçi Historic Urban Landscape (HUL). The study first discusses the repair, maintenance, and restoration process, the different actors involved in the process, and the related legal frameworks in Turkey. In the following section, the status of the current housing stock in the Kılınçaslan neighbourhood is determined, also detailing the changes in the number of housing units in Kaleiçi over time and the applications made for interventions to the Regional Conservation Council for Kaleiçi. Following this, the findings of site surveys of housings units and vacant buildings in the neighbourhood detailing their current status are discussed, and the applications for interventions made to KUDEB are then analyzed. In the final part, policies are suggested related to the maintenance of the current traditional housing stock and their current residents.

Legal Framework Related to the Repair, Maintenance, and Restoration of Traditional Buildings, and the Availability of Grants

There are numerous actors involved in the repair, maintenance, and restoration of listed historic buildings in Turkey, including the Ministry of Culture and Tourism, Regional Conservation Councils, TOKİ, KUDEB, municipalities, and residents. While some of these actors are involved in the monitoring and management of such implementations, including KUDEB and Regional Conservation Councils, others provide grants for the repair, maintenance, and restoration of such buildings.

The Ministry of Culture and Tourism is the main actor responsible for the provision of grants to owners of listed historic buildings, contributing to the budget of restoration projects and implementations. The Law on the Conservation of Cultural and Natural Property (Law No. 2863), in the section on Assisting and Contributing to the Repair of Immovable Cultural Properties Part of Article 12, states that the Ministry provides in-kind, cash, and technical assistance for the protection, maintenance and repair of the cultural and natural properties owned by real and legal persons, and are subject to the Civil Law. Grants are given both for projects and implementations. For example, in İstanbul in 2020, one in every nine project applications resulted in a grant, along with two of the six implementation applications (Akçaöz, 2021). In 2021, five project applications and four implementation applications were made, which is very low considering the rich traditional building stock in need of restoration in İstanbul (Parlak, 2021). Regional Conservation Councils decide upon the construction and physical interventions, as well as implementations other than repairs and changes that do not require a permit on the sites of registered immovable cultural property, as indicated in the Regulation of the High Council on the Conservation of Cultural Properties and the Regional Conservation Council (Article 11 and Item ğ). The restoration approval and supervision of such processes are coordinated by these councils, and project preparations take roughly three to six months, although the process may be longer. The granting of approval of projects may also take a long time (Pekol and Kayasü, n.d.), which may place residents who need to maintain their houses in a difficult situation.

KUDEBs are established within metropolitan municipalities, special provincial administrations, and municipalities upon the agreement of the Ministry of Culture and Tourism. Although they do not directly finance repairs, maintenance, and restorations, they supervise the conservation process of immovable cultural property. Carrying out assessments of buildings prior to repair or maintenance works, regulating the preliminary permission documents, providing a certificate of conformity for repairs, and ensuring repairs and maintenance are authentic in form and the materials used in registered buildings are under the responsibility of KUDEB. During maintenance and repair works, if it is found that major repairs[3] are required, KUDEB will call a halt to the maintenance and repair and report the issue to the Regional Conservation Council. KUDEB's establishment regulations it is emphasized that the two bodies are to arrange grants with the relevant institution for owners who lack the necessary resources for the repair of immovable cultural property (Article 7, Item i). They are involved directly as technical staff, or supervise those who take on this duty on behalf of owners who are unable to repair the cultural property (Article 7, Item j). To apply for an intervention, the resident writes an application letter to KUDEB, after which a member of staff of KUDEB assesses the situation on the site. After preparing the necessary documents for issuance of a Certificate of Preliminary Permission for Repair, the resident enters into a deal with an architect or civil engineer who will be responsible for the repair. A Certificate of Conformity for Repair is granted if the suggested intervention is found suitable, and is valid for one year. Such an application and implementation process may take several months (Pekol and Kayasü, n.d.: 17), and may thus contribute to the further deterioration of the building, since simple repairs include such urgent interventions as roof repairs. During a discussion with two residents in Kaleiçi during the site surveys, this bureaucratic process was highlighted as a challenge.

TOKİ also provides grants, in accordance with Law No. 2863, as detailed in the "Assisting and Contributing to the Repair of Immovable Cultural Properties" part of Article 12. According to the Mass Housing Law No. 2985, at least 10 percent of the loans to be granted are to be used for implementations involving the maintenance, repair, and restoration of registered immovable cultural properties that are the property of real and legal persons subject to private law (TOKİ, n.d.). The priority projects are determined by the Ministry and TOKİ, with a maximum of 70 percent of the estimated cost of the project coverable by the loan. After the necessary guarantees have been received by the relevant bank, 15 percent of the loan is paid in advance (TOKİ, n.d.).

Furthermore, in accordance with Regulation on the Share of Contributions for the Conservation of Immovable Cultural Properties, municipalities and special provincial administrations are to use 10 percent of the collected property taxes for the conservation of immovable cultural properties located within their areas of responsibility. Provided as grants, these funds can be used for the drawing up of plans, projects, implementations, and expropriations by special provincial administrations and municipalities, under the control of the local governor, while special provincial administrations supervise the distribution process. No more than 30 percent of the sum allocated for the province can be used for projects of the special provincial administration. Other forms of direct financial support include expropriations by the Ministry of Culture and Tourism (purchasing and bartering) and the transfer of development rights, while indirect forms of support include exemptions from income and corporation tax for donations benefiting cultural and natural properties, and exemptions of inheritance and transition taxes and reduction in income and payroll taxes (Ulusan and Ersoy, 2018: 254). These sources, however, were not examined in detail in the present study, as they do not serve as a source of grants for residents.

The high costs associated with rehabilitation works and the strict conservation rules result in unaffordable situations for the long-term residents of historical buildings, and in the event of there being a lack of resources, local authorities may expropriate traditional housing stock while investing in the upgrading of public property (Serageldin, 2000: 54). As stated by Şahin Güçhan and Kurul (2009), despite the presence of a comprehensive legal structure in Turkey, the implementation of key decisions is very difficult due to the limited financial sources. As Ulusan and Ersoy (2018: 264) claim, the financial support provided for the conservation of registered historic buildings covers only 0.66 percent of the estimated total cost. Considering the large number of registered historic buildings in Turkey—almost 73,000 registered at the end of 2020 (KVMGM, n.d.) —these funding sources can be considered very limited.

The residents of housing units are involved both as investors and as coordinators of the repair, maintenance, and restoration process, as stated in Law No. 2863, Article 11. In the same Law, it was also emphasized that if the owners of properties are unable to fulfil their maintenance and repair responsibilities, their properties will be expropriated. Given the desire to ensure the continued authentic use of traditional housing by their occupants, this situation could be problematic, since not every owner is able to afford such costs. Furthermore, the complicated procedures to be followed related to the repair, maintenance, or restoration of such properties places an additional burden on the residents of housing units. The legal framework falls short of clearly defining the restoration process, the actors and their positions, the authorities and the assignment of responsibilities, and furthermore, the survey techniques, restitution studies or restoration decisions, and even the project phase, are full of ambiguities (Kaynak, 2016).

It is also unclear under what conditions owners may apply for different grants. For example, in the grants provided by TOKİ, priority is given to local administration projects (Yetgin, 2005). An inability to access sufficient funding may compel owners to abandon the building, or force them to live in substandard conditions. Furthermore, the residents of registered buildings lack the freedom to make necessary interventions, and this situation leads to a decrease in the useable housing stock, and again, may result in them living in substandard conditions. Accordingly, the number of vacant units in traditional urban areas continues to rise.

Research Method

The focus of the present study is the Kılınçaslan neighbourhood in the Antalya Kaleiçi HUL, which is made up of four neighbourhoods, namely Selçuk, Tuzcular, Barbaros, and Kılınçaslan. Among these, the Tuzcular and Selçuk neighbourhoods have been transformed for commercial use over time due to their direct interactions with the current city centre. The Barbaros neighbourhood contains fewer housing units than the Kılınçaslan neighbourhood, and for this reason, the Kılınçaslan neighbourhood is selected for scrutiny in the present study, having the largest population and the largest number of housing units in the Kaleiçi HUL. Data for the research were garnered from site surveys, a literature review and stakeholder institutions.

The status of the current housing stock in the Kılınçaslan neighbourhood was ascertained from two[4] site surveys. The current condition of the housing stock was assessed as structural stability is one of the leading issues in the extensive surveys related to the conservation of cultural properties (Umar, 2019). Accordingly, building conditions are classified based on a visual inspection of their exterior to identify any structural problems, surface deterioration, material loss, and decay. However, further assessments should be made of the interior of historic buildings to ascertain the overall condition. Ruined buildings and those undergoing restoration and construction were also noted in the neighbourhood. The exterior condition of the housing units and vacant buildings were classified according to the following criteria:

Buildings in Good Condition: No structural problems are apparent in the facades of building.[5]

Buildings in an Adequate Condition: No structural problems, but minor problems apparent on the facade and some surface deterioration.

Buildings in Poor Condition: Structural problems, material loss and decay to the facades.

Ruined Buildings: Buildings that have suffered considerable destruction.

Buildings in Restoration Process: Buildings that undergoing restoration.

Buildings in Construction Process: Buildings that are undergoing construction with concrete systems.

Housing units and vacant buildings that could function as housing were identified in the neighbourhood, and were categorized with respect to their exterior condition and use. Following the site surveys, the buildings were categorized as "in good condition and in use", "in an adequate condition and in use", "in a poor condition and in use", "in good condition and vacant", and "in a poor condition and vacant". No buildings in the neighbourhood were considered to be "in an adequate condition and vacant" during site surveys.

Data on the buildings granted Preliminary Permission Certificates for Repair and Certificates of Conformity for Repair in Antalya were taken from the Activity Report of Antalya Metropolitan Municipality (2013, 2014, 2015, 2016, 2017, 2018, 2019) and a Preliminary Certificate of Repair in Kılınçaslan Neighbourhood was obtained from KUDEB. Furthermore, details of applications made for different interventions in Kaleiçi were garnered from the meeting minutes of the Antalya Conservation Council, for which 78 meeting minutes were analyzed.

Findings and Discussions

Kaleiçi and its Housing Stock

Antalya Kaleiçi, which is a multi-layered historic urban landscape located on the hills that line the south of the city has been host to different historic settlements in time, including those of the Roman, Byzantine, Seljuk, Ottoman, and Republican periods. Today, the urban pattern and the traditional housing units are mostly from the Ottoman period. The Kılınçaslan neighbourhood is located in the southeast of Kaleiçi (Figure 12.1), which has a direct physical, social and economic relationship with the current city centre, and is at the core of the city centre. It serves as a hub for different public transport modes and is close to many important urban facilities, such as Antalya International Airport, the Intercity Bus Station, and Akdeniz University. As seen in Figure 12.1, it has in its close surroundings the Balbey and Haşim İşçan Urban Sites, Kapalı Street, İnönü Square, Cumhuriyet Square, and commercial areas that are made use of by Antalya residents, as well as local and foreign tourists. Its relationship with these uses has affected its transformation from residential use into a touristic area, while its location on the hills and its direct relationship with the Mediterranean Sea increase its attractiveness (Figure 12.2).

The first conservation and development plan for the Kaleiçi area – the 1979 Kaleiçi Conservation Plan- was made when it was a common residential neighbourhood. As a result of the conservation, refunctioning and revitalization strategies laid out in the plan, Kaleiçi became an important district for tourism, trade, and entertainment, and the old port was transformed into a marina (Kamacı and Örmecioğlu, 2005). Following the transformation, many residents opted to sell their houses (Özgönül, 2015), and buildings

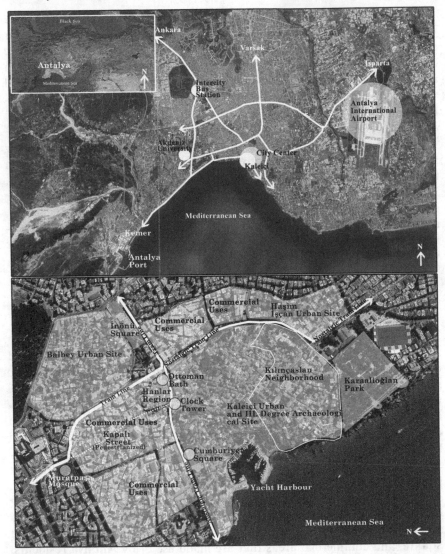

Figure 12.1 Location of the study area and its relationship within the city.

Source: Prepared by the author on base maps taken from Google Maps and Google Earth.

serving touristic and cultural purposes thus began to increase day-by-day in the region where the use of traditional housing was intense. Many historic buildings have been transformed into hotels, museums, or cafes (Uslu, 2018a: 874), and today, Kaleiçi features a diversity of land uses, including pensions, hotels, mosques, education establishments, housing, cafes, and many other commercial facilities serve the tourism market. The transformation of the

Figure 12.2 View of Kaleiçi.

Source: Uluç's Personal Archive.

Kaleiçi HUL into a tourism destination has resulted in a gradual decrease in the number of housing units in time. According to a study by Öztekin (2010), the number of housing units in Kaleiçi has decreased sharply from year to year, from 758 in 1945, to 781 in 1979 and 140 in 2009 (Öztekin, 2010: 87), and rising to 202 by 2018 (Figure 12.3[6]). In contrast, the number of touristic facilities has increased gradually over time, from 2 in 1979 to 145 in 2009 (Öztekin, 2010).

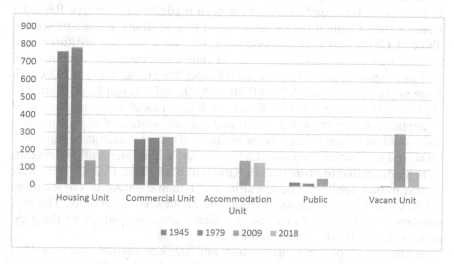

Figure 12.3 Changes in Kaleiçi housing, commercial, accommodation, public, and vacant units.

Source: This table is compiled from different data. 1945, 1979, and 2009 data were taken from Öztekin (2010) research. Data from 2018 were taken from a land-use map. The units on each plot were counted taking into account their ground floor function.

Table 12.1 Different applications made to the Antalya Regional Conservation Council in 2019 and 2020

Application related to the Kaleiçi Urban and the 3rd Degree Archaeological Site made to Antalya Conservation Council	2019	2020
Restoration	14	9
Reconstruction	2	1
Repair	4	4
Survey	3	1
Restitution	3	–
Survey, restitution, restoration	2	1
Restoration and change in function	1	1
New construction	3	2
Change in function	8	–
Occupancy permit	2	1
Building registration certificate	37	5
Total	**79**	**25**

Source: Prepared by the author based on Regional Conservation Council Meeting Minutes published in https://korumakurullari.ktb.gov.tr/.

In Kaleiçi today, almost the entire site is taken up by touristic facilities and commercial activity. As can be understood from a 2018 land-use map, there were around 88 vacant and 30 ruined buildings in Kaleiçi.

Kaleiçi's authenticity, integrity, and strategic location are highly dynamic in terms of both the planned and completed interventions and the changes in land use. Many applications have been made to the Antalya Regional Conservation Council related to different interventions. As can be seen in Table 12.1,[78] the applications made for restorations, surveys, restitution and repairs, new constructions, changes in function, occupancy permits, and building registration certificates in 2019 differ from those made in 2020, with more applications made in 2019 than in 2020, and this can be attributed to the Covid-19 pandemic declared in 2020. It is not clear, however, what kind of applications are made for traditional housing units in Kaleiçi, although they could include applications for buildings used for touristic, commercial, civic, and residential purposes. The figures show that intervention demands were made related to properties in Kaleiçi in both years. In 2019, there were 14 restoration applications, revealing the dynamic nature of the location, while there were also applications made for new constructions. The new construction applications can be considered controversial, given the presence of stock with the potential for use.

In the report of the Conservation and Implementation Revision Plan (Anonymous, 2018), the requested changes in building function were as follows: 8.3 percent of those being used as workplaces were requested to be transformed into housing units; 41.7 percent were to be transformed into housing or workplaces; and 50 percent were being used as workplaces, but were requested to be changed into a different type of workplace. The requests made to refunction buildings for residential use are of particular interest, as a large proportion of the applications seek to refunction the building as combined housing and workplaces (Anonymous, 2018). It was also identified in the present study that owners are more likely to spend money on repairs than tenants, with 72.1 percent of the users making expenditures related to the building being the owners, with the remaining 27.9 percent being tenants (Anonymous, 2018). Accordingly, as can be expected, owners are more willing to spend money on their properties than tenants. Also, housing demand exists for Kaleiçi by current users.

Housing Stock in Kılınçaslan Neighbourhood

An analysis of the number of households reveals no significant difference in household numbers in the Kılınçaslan neighbourhood between 2012 and 2018 (Table 12.2), although an increase in household numbers was noted after 2015.

As discussed previously, Kaleiçi is a highly dynamic site in terms of land-use changes and requests for interventions made to the Regional Conservation Council and KUDEB. The number of vacant buildings is particularly worthy of note, as there is potential for them to be used as housing units, but also to be left to ruin. The buildings being used for housing and the vacant properties were examined during the site surveys in the present study. Both traditional and new residential buildings are detailed on the below map, along with registered buildings and housing units whose ground floors have different functions (Figure 12.4).[9]

As can be seen from Figure 12.4, 76 buildings are used for housing in the neighbourhood, and the ground floor of four of these have other functions. There are 23 vacant buildings in the neighbourhood, 11 of which are registered, while 28 of the total 76 housing units that are in use are registered. The condition of the vacant buildings and housing units was assessed and classified based on the results of the site survey. The condition of the 76 units, some of which are in use, while others are vacant, was assessed and categorized (Table 12.3). At first, three basic condition groups were used to assess the condition of the vacant buildings and housing units, being good, adequate, and poor; however, during the site surveys, the ruined buildings, those undergoing restoration and those under construction process were also explored (Figure 12.5).

Table 12.2 The number of households between 2012 and 2018 in the Kılınçaslan neighbourhood

2012	2013	2014	2015	2016	2017	2018
189	191	183	186	189	199	204

Source: TURKSTAT (2018).

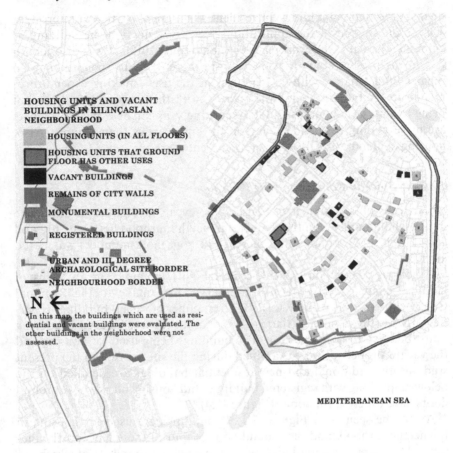

Figure 12.4 Housing units and vacant buildings in Kılınçaslan neighbourhood.

Source: Prepared by the author based on data gathered during the site surveys.

Table 12.3 The number of vacant and in-use housing units and their conditions

Condition ╲ status	Vacant buildings	Housing units in use
Good	9	50
Adequate	0	19
Poor	6	7
Ruin	2	0
In restoration	5	0
Under construction	1	0
Total	**23**	**76**

Source: Prepared by the author based on data gathered during the site surveys in 2020.

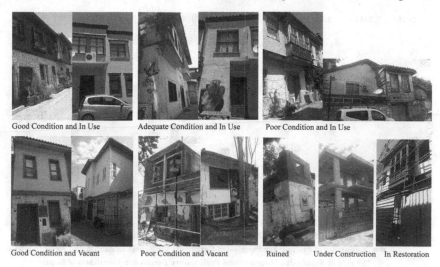

Good Condition and In Use   Adequate Condition and In Use   Poor Condition and In Use

Good Condition and Vacant   Poor Condition and Vacant   Ruined   Under Construction   In Restoration

Figure 12.5 Different housing units investigated in the Kılınçaslan neighbourhood.

Source: All photos taken by the author during site surveys conducted in 2020.

The housing units considered to be in a good condition and in use (50 buildings) look good from the exterior; they have no apparent structural problems and the external render appears to be in good condition. The buildings in an adequate condition and in use (19 buildings) appear to have no structural problems, although some surface deterioration is apparent. Among the vacant buildings, nine appear to be in good condition. Most of the buildings (69 buildings) that are in use are in a good or adequate condition, while six vacant buildings were assessed as being in poor condition, with deteriorated rendering, and are in need of immediate restoration. It is clear that if disinvestment continues, these buildings will eventually end up being demolished. The survey identified two ruined traditional buildings in the neighbourhood, one building that was under construction and five that were undergoing restoration.

It should be emphasized here that vacant buildings in a good or adequate condition may be used for housing, although it is not clear whether those identified in the study will be used for this purpose. As can be understood from an analysis of the dynamic touristic side of the site, they could be refunctioned for other uses. It is uncertain whether or not the buildings undergoing restoration or construction will be used as housing or for other purposes, although they can be considered as a potential housing source in the location.

For simple repairs, applications are made to KUDEB, since even the smallest rehabilitation and maintenance to be conducted in Kaleiçi requires approval (Uslu, 2018a: 874). Figure 12.6[10] shows the number of Preliminary Permission Certificates for Repair, the Conformity Certificates for Repair

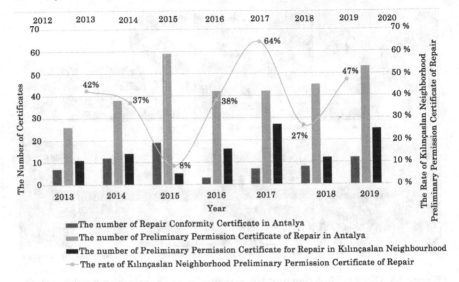

Figure 12.6 The number of preliminary permission certificates of repair and repair conformity certificates in Antalya and the Kılınçaslan Neighbourhood.

Source: Prepared by the author based on Activity Reports of Antalya Metropolitan Municipality and KUDEB data.

issued in Antalya, and the number of Preliminary Permission Certificates for Repairs in the Kılınçaslan neighbourhood. The number of Preliminary Permission Certificate for Repair and the number of Repair Conformity Certificates in Antalya increased gradually between 2013 and 2019.

Between 2013 and 2019, 110 Preliminary Permissions for Repair were granted in the Kılınçaslan neighbourhood. Looking at the rate of Preliminary Permission Certificates for Repair issued for the Kılınçaslan neighbourhood, it can be seen that significant applications are made each year. In 2017, this rate reached 64 percent. This reveals the dynamism of the Kılınçaslan neighbourhood in terms of the applications made for repair. It is not clear, however, how many applications are made specifically for housing units in the neighbourhood, as these applications could have been made for other uses. If they were made for interventions to upgrade commercial or accommodation units, it could be said that residents are unable to afford repair costs. As discussed with two residents during the site surveys, the difficulty in obtaining permission for simple but necessary interventions and the high costs of restoration were highlighted.

Conclusion

There are different challenges faced by the residents of registered housing units in Turkey. First, living in a registered building comes with limitations related to interventions into properties, as the development rights of owners of registered buildings are controlled by laws and regulations. In addition, although these buildings may have various special values, the responsibility for maintenance, simple repairs, and restoration is placed on the residents in the legal regulations, as is the case with private housing stock. Furthermore, obtaining financing for the repair, maintenance and restoration can be difficult due to the high costs involved, and the public grants and loans provided are insufficient, while the long bureaucratic process further complicates the repair, maintenance, and restoration processes, meaning that the occupants must live in substandard conditions. The processes involved prior to interventions include gaining the permission and approval of KUDEB and Regional Conservation Councils, and these processes can take time to complete. Such challenges could lead to a decrease in residential use and the abandonment of properties, as well as them being left in a state of poor repair or occupation under poor conditions. Accordingly, the presence of vacant buildings and those in poor condition in traditional areas is inevitable, and the associated high vacancy rates could lead to such urban areas being lost.

The number of registered buildings varies from city to city. Some cities have huge areas of traditional housing that have their own specific spatial organizations, technical infrastructures, and social networks. If existing housing stock is not used, areas containing a large proportion of vacant housing can suffer from areal vacancy, and the associated problems of security, arson, and urban deterioration in the city centres that arise in time. Besides, when the number of households is fixed, the presence of vacant housing units that are not evaluated or used leads to a need for new housing provision, leading to new housing construction.

This case study of Antalya Kaleiçi has shown that the number of in-use housing units decreases gradually from year to year. In addition to the current legal regulations limitations and challenges, Kaleiçi's position in the city centre has contributed to this situation, given its direct interaction with the current city centre, the public transport system, the pedestrianized Kapalı Street, the Mediterranean Sea, and the Haşim İşçan and Balbey urban areas. An analysis of the intervention applications made to the Regional Conservation Council and KUDEB allows the dynamic side of Kaleiçi to be assessed. Although it is not clear for what kinds of buildings the applications are made, there is a clear indication of the interventions required to be conducted. The site survey identified five housing categories in the Kılınçaslan neighbourhood in Kaleiçi: the first of these categories is housing that is in good condition and in use, which can be found distributed throughout the neighbourhood, not all of which are traditional, having been constructed with modern construction systems. The second category is houses that are in an adequate condition and that are in use. These buildings have shown signs of deterioration, and are spread throughout the neighbourhood. The third category is houses that are in a poor condition and that are in use.

These units are in need of restoration, and their continued use as substandard accommodation may create risks to the resident households. The fourth category includes housing units that are in good condition and vacant, some of which have been restored. The fifth and final categories are housing units that are in a poor condition and vacant. These buildings have structural problems, and have suffered material loss and decay to the facades, and are in need of urgent comprehensive restoration projects if they are to be saved from demolition. The buildings that are in a poor condition, whether they are vacant or in use, need urgent interventions, however, the high costs of restoration and the bureaucratic hurdles bring challenges to this situation. In recognition of the increase in household numbers between 2012 and 2018, the surveys made for the Conservation Development Revision Plan have identified a need to transform workplaces into housing units.

Taking the above into account, there is a need for policies that support the repair, maintenance, and restoration processes for residents. From a conservation perspective, the conservation and development of heritage sites could be maintained by ensuring the continued residence of the host communities and their use of historic buildings. Firstly, for repair, maintenance, and restoration activities, financial support should be provided to the residents. While some direct and indirect financial resources are detailed in the current legal and administrative regulations in Turkey for the repair, maintenance, and restoration of traditional housing units, there is a need for further support. In this regard, cooperation with the private sector could be considered as an alternative approach to the conservation of heritage, as has been emphasized in various international documents (Madran and Özgönül, 1999 cited in Ulusan and Ersoy, 2018: 265). Furthermore, the collaborations of households and administrative bodies should be improved. Although KUDEB contributes to this process, the involvement of residents should be increased, and their problems and needs should be taken into account in more basic processes. Furthermore, the application and approval processes related to the repairs needed by residents should not put the residents in difficult situations, and so the current long bureaucratic process should be shortened so that residents do not see such procedures as a burden.

In conclusion, making use of traditional housing stock is essential, as is ensuring the authentic residential use of such structures by their residents. Using existing housing stock rather than constructing new housing is economically preferable, and prevents the transformation of traditional urban areas into deteriorated areas. In such situations, providing households with the necessary financial tools gains importance. Finding a balance between the needs of the current residents of traditional housing units and the legal regulations and limitations should be considered a priority. It can be said that the economic and technical problems faced by the residents of heritage sites related to restoration processes are common in different country contexts. Future studies could look at similar situations in different countries, and the different perspectives could be discussed so as to contribute to a better understanding of the problems experienced in traditional housing areas.

Acknowledgements

I give thanks to Ali Kahraman, the mukhtar of the Kılınçaslan neighbourhood, for accompanying me during the first site survey. Without his guidance, the identification of the housing units on the site would have been challenging. I give special thanks also to the editors and the anonymous referees for their constructive comments on the previous versions of the study.

Notes

1 In this study, the concept of historic building is mainly used for civil architecture examples.
2 In the Turkey context, restorations requiring extensive interventions require architectural projects, while maintenance and simple repairs do not. Maintenance refers to such interventions related to roof, gutter repairs, and rendering works that are aimed at prolonging the life of building and that do not require changes in design, material, structure, or architectural elements. Simple repairs include the replacement of decaying architectural elements, such as wood, metal, terracotta, or stone, with the same material, in their original form and replacing damaged interior and exterior plasters and coatings that are harmonious in terms of color and material.
3 Major repairs include implementations based on survey, restitution, and restoration projects that have been prepared according to scientific principles, other than maintenance or repairs.
4 The first site survey was conducted on February 13 and 14, 2020, and investigated the housing units and vacant buildings in the neighborhood. The second site survey was conducted on May 30 and 31, 2020 and assessed the condition of buildings, and the findings were used in the revision of the first site survey data.
5 Structural problems refer to cracks and decay to columns and beams, and large-scale cracks and deteriorations to roofs.
6 This table was compiled based on different data. Data for 1945, 1979, and 2009 were taken from the study by Öztekin (2010), while 2018 data were obtained from a land-use map from the said year. The units in each plot were counted with respect to their ground-floor use.
7 The Minutes of 78 Meetings of the Antalya Conservation Council, published at (https://korumakurullari.ktb.gov.tr/) were analyzed. The meeting minutes provided details of repairs, maintenance, surveys, restitutions, and restorations in the Antalya Kaleiçi Urban and 3rd Degree Archaeological Site.
8 Definitions based on the Regulation on Cultural Property Tenders.
The restitution project includes drawings related to the analysis of cultural properties and their close surroundings.
The restoration project covers intervention techniques for the original and suggested functions of cultural properties.
The survey project includes the report and scaled project of the current status of cultural properties and their close surroundings.
9 Registered building data and the Urban and 3rd Degree Archaeological Site border were obtained from the Antalya Metropolitan Municipality.

10 Data related to the Preliminary Certificate of Repair and the Repair Conformity Certificate in Antalya were obtained from the Activity Reports of the Antalya Metropolitan Municipality, while data related to the Kılınçaslan neighborhood were obtained from KUDEB.

References

Activity Reports of Antalya Metropolitan Municipality (2013, 2014, 2015, 2016, 2017, 2018, 2019). Available at: https://www.antalya.bel.tr/. Accessed 19 March 2021.

Akçaöz M (2021) Restorasyon için Finansal Destekler-KMKD. Available at: https://www.youtube.com/watch?v=fd6_P9Zrkrk&list=PLrdtmupiGAovfK5X2vxrqy JO5lxde68r6&index=5. Accessed 19 March 2021.

Anonymous (2018) Antalya Büyükşehir Belediyesi Antalya Kent Merkezi Kültür ve Turizm Koruma ve Gelişim Bölgesi Muratpaşa İlçesi Kaleiçi Kentsel ve III. Derece Arkeolojik Sit Alanı 1/1000 Ölçekli Koruma Amaçlı Uygulama İmar Planı Revizyonu Plan Açıklama Raporu.

Arabacıoğlu FP, Aydemir I (2007) Tarihi çevrelerde yeniden değerlendirme kavramı. *Megaron* 2(4): 204–212.

Carter B, Grimwade G (1997) Balancing use and preservation in cultural heritage management. *International Journal of Heritage Studies* 3(1): 45–53.

CoE (1975) *The Declaration of Amsterdam*. Amsterdam: CoE.

Ecemiş Kılıç S, Türkoğlu G (2015) Conservation problems of traditional housing with continued original function and recommended solutions: Safranbolu "Eski Çarşı District". *Megaron* 10(2): 260–270.

Elsorady DA (2011) Heritage conservation in Alexandria, Egypt: managing tensions between ownership and legislation. *International Journal of Heritage Studies* 17(5): 497–513.

Halifeoğlu FM, Işık N, Tekin M et al. (2020) Geleneksel mimaride yeniden kullanım kaynaklı sorunlar: Diyarbakır evleri örneklemesi. *Uluslararası Sosyal Araştırmalar Dergisi* 13(75): 382–395.

Historic England (2011) Stopping the rot: A guide to enforcement action to save historic buildings. Available at: https://historicengland.org.uk/images-books/publications/stoppingtherot/heag046b-stopping-the-rot/. Accessed 16 April 2021.

ICOMOS (1987) *Charter for the conservation of historic towns and urban areas (Washington Charter 1987)*. Washington, DC: ICOMOS.

ICOMOS (2011) *The Valletta principles for the safeguarding and management of historic cities, towns and urban areas*. Paris: ICOMOS.

Jahed N, Aktaş YD, Rickaby P et al. (2020) Policy framework for energy retrofitting of built heritage: a critical comparison of UK and Turkey. *Atmosphere* 11(674): 1–19.

Kamacı E, Örmecioğlu HT (2005) *Kentsel sit alanlarında kentsel dönüşüm problematiği: Antalya Kaleiçi*. Available at: https://www.imo.org.tr/resimler/ekutuphane/pdf/11172.pdf. Accessed 8 March 2021.

Kaynak S (2016) *Eski eser restorasyon proje süreci: sorunlar ve çözüm önerileri*. Available at: http://www.konakmimarlik.com/eski-eser-restorasyon-proje-sureci/. Accessed 16 March 2021.

KVMGM (n.d.) Türkiye Geneli Korunması Gerekli Taşınmaz Kültür Varlığı İstatistiği. Available at: https://kvmgm.ktb.gov.tr/TR-44798/turkiye-geneli-korunmasi-gerekli-tasinmaz-kultur-varlig-.html. Accessed 21 March 2021.

Madran E, Özgönül N (eds) (1999) *International Documents Regarding the Preservation of Cultural and Natural Heritage.* Ankara: METU Faculty of Architecture Press.

Madran E, Özgönül N (2005) *Kültürel ve Doğal Değerlerin Korunması.* Ankara: TMMOB Mimarlar Odası.

Medici SD, Bellomia M and Senia C (2019) Adaptive reuse and rehabilitation of cultural heritage. A performance-based approach for fire safety. Available at: http:// www.classiconorroena.unina.it/index.php/bdc/article/view/7062/8004. Accessed 26 March 2021.

Mısırlısoy D, Günçe K (2016) A critical look to the adaptive reuse of traditional urban houses in the Walled City of Nicosia. *Journal of Architectural Conservation* 22(2): 149–166.

Özgönül N (2015) Turkish involvement in the 1975 European heritage year campaign and its impact on heritage conservation in Turkey. In: Falser M, Lipp W (eds). *A Future for Our Past: the 40th Anniversary of European Architectural Heritage Year (1975–2015).* Osterreich: ICOMOS, 332–345.

Öztekin D (2010) *Sosyal ve fiziksel çevre bağlamında koruma planları Antalya Kaleiçi örneği.* İstanbul: YTÜ Fen Bilimleri Enstitüsü.

Parlak Ç (2021) Restorasyon için finansal destekler-KMKD. Available at: https:// www.youtube.com/watch?v=fd6_P9Zrkrk&list=PLrdtmupiGAovfK5X2vxrqy JO5lxde68r6&index=5. Accessed 19 March 2021.

Pekol B, Kayasü S (n.d.) Tarihi Evlerde Yaşam Tarihi Ev Kullanıcısının El Kitabı. Available at: http://kmkd.org/mimari-koruma-el-kitaplari/. Accessed 30 March 2021.

Saka AE, Özen H (2013) Tarihi yapılarda yeniden kullanım sorunları Tokat Meydan ve Sulu Sokak. *Sosyal Bilimler Araştırmaları Dergisi* 8(1): 23–48.

Serageldin M (2000) Preserving the historic urban fabric in a context of a fast-paced change. In: Avrami E, Mason R, De La Torre M (eds) *Values and Heritage Conservation.* Los Angeles: The Getty Conservation Institution, 51–58.

Stovel H (2002) Approach to managing urban transformation for historic cities. *Review of Culture* 4: 35–44.

Şahin E (2006) *Evaluation of financial instruments within the conservation activities,* Unpublished master thesis. Middle East Technical University.

Şahin Güçhan N, Kurul E (2009) A history of the development of conservation measures in Turkey: From the mid 19th century until 2004. *METU Journal of the Faculty of Architecture* 26(2): 19–44.

TOKİ (n.d.) Restorasyon Kredileri. Available at: https://www.toki.gov.tr/restorasyon-kredileri. Accessed 21 March 2021.

TURKSTAT (2018) The number of households from address based population registration system (ADNKS).

Ulusan E, Ersoy M (2018) Financing the preservation of historical buildings in Turkey. *METU Journal of the Faculty of Architecture* 35(2): 251–267.

Umar GK (2019) Methodological issues in architectural conservation, preservation and restoration of Hausa traditional residential building: Case study of Kano Metropolis. *International Journal of Advanced Academic Research* 5(2): 104–113.

Uslu F (2018a) 1960 sonrası turizm ile birlikte gelen sosyo-kültürel ve mekansal değişim: Antalya Kaleiçi örneği. *Turkish Studies* 13(3): 865–887.

Uslu F (2018b) Bin yıllık Türk yurdu Antalya Kaleiçi'nin son yıllarda turizmle birlikte oluşan sosyo-kültürel değişimi. *Atatürk Üniversitesi Sosyal Bilimler Enstitüsü Dergisi* 22: 811–831.

Wang J (2012) Problems and solutions in the protection of historical urban areas. *Frontiers of Architectural Research* 1(1): 40–43.

Yetgin F (2005) Tarihi kent merkezlerinin yenilenmesinde inşaat sektörünün rolü ve yeri. In: Özden P (ed) *Tarihi Kent Merkezlerinde Yenileme: Beklentiler, Sorunlar, Uygulamalar*. İstanbul: TMMOB Şehir Plancıları Odası.

Yung EHK, Chan EHW (2012) Implementation challenges to the adaptive reuse of heritage buildings: Towards the goals of sustainable, low carbon cities. *Habitat International* 36(3): 352–361.

Part IV

Housing Policy (or Lack Thereof) and Its Implications

13 Housing for Ageing Populations in Turkey

Yelda Kızıldağ Özdemirli

Introduction

The demographic trend of ageing is not unique to Turkey but a universal one. The share of individuals aged 65 or over has increased globally, especially after the 1990s (United Nations, 2019). There was only one person aged 65 or over in every sixteen individuals in the 1990s. This proportion rose to one in every eleven in 2019, and it is projected that by 2050, one in six people will be aged 65 years or over worldwide.

In Turkey, there are nearly 7.6 million individuals in the 65 years or over age group, which constitutes around 9.1 per cent of the population as of 2019 (Figure 13.1). This figure has been achieved by a 21.9 per cent increase rate in the last five years. According to the projections of the Turkish Statistical Institute (TURKSTAT), the rate of people 65 years or over will be 16 per cent in 20 years and 26 per cent in 40 years, depicting a very rapid increase. Thus, Turkey will be among the aged populations by the middle of the century. Evidence from the population ageing experience of other parts and regions of the world such as Europe indicates that this trend might lead to many challenges to the elderly and society in general in terms of housing provision, satisfaction, and affordability (Houben, 2001; Sixsmith et al., 2014; Abramsson and Andersson, 2016; Herbers and Mulder, 2017; European Commission, 2019).

Although there is not a consensus on the definition of the "old population", according to most countries' statistics, old age begins at 60 or 65 years of age. Some of the United Nations (UN, 2002, 2010, 2015) publications assume age 60 and above as old age. World Health Organization (WHO, 2011, 2015) defines 65 and older as "old". Traditionally, "old" is categorized into subgroups; the young-old (ages 65–74), the middle-old (ages 75–84), and the oldest-old (over age 85). TURKSTAT also accepts 65 and older as old and defines this age group as a "dependent" population in economic terms as well. Besides chronicle age, "old age" is a socially constructed term, mostly indicating that active contribution to the economy is no longer possible and 60 and 65 years of age are roughly equivalent to retirement ages in most parts of the world. Correspondingly, in this study, the terms old, elderly and ageing populations will be used interchangeably to refer to those aged 60 or 65 years or over.

DOI: 10.4324/9781003173670-17

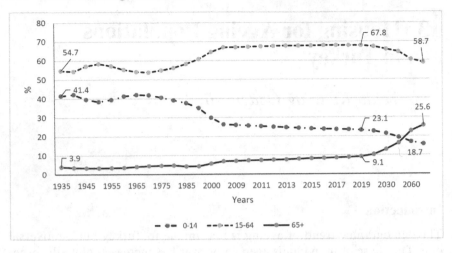

Figure 13.1 Change in broad age group proportions in Turkey between 1935 and 2060.

Source: Derived by the Author from Elderly Statistics, 2019 (TURKSTAT, 2019a).

Being an inevitable phase of life, old age is a concern for society, cutting across all social, economic, gender, ethnic, (dis)ability, educational background, and occupational lines. Although housing policies for particular groups, such as low-income populations, exist in Turkey, the housing needs of the elderly have hardly been acknowledged, adequately addressed, or prioritized. There are at least two major reasons why housing for ageing populations has not attracted enough attention in the housing policy of Turkey. Firstly, Turkey has had a fairly young population structure, with relatively high population growth rates, resulting in older populations being overlooked. Secondly, especially after the 1950s, population growth, inner migration from rural to urban areas, and more recently, international migration all have emphasized an urgent need for housing provision. This urgency resulted in favouring quantity over quality, lacking also responsiveness to some of the socio-demographic attributes and trends. This study aims to highlight this huge gap in the literature, policy, and practice realms by putting the housing need of the ageing section of the population into perspective through the analysis of the current situation of demography, housing stock, and housing policy. Correspondingly, the study has been designed to explore three main research questions:

1   What are the housing needs of the older households and how are they different from other age groups?
2   What is the current capacity of the housing stock and urban amenities in Turkey in terms of meeting these needs?
3   What are the available tools and potentials of the housing policy in Turkey to meet the needs of the ageing population?

The research methodology includes the quantitative analysis of country-level demographic data, the analysis of housing-related data collected by various institutions (including TURKSTAT, UN, and World Bank), literature research, observation, archival research on relevant media coverage, and the reports of relevant government institutions. Moreover, examples of elderly housing policy and practice throughout the world are briefly examined to discuss possible routes for housing policy in Turkey.

The first part of the chapter reviews the current demographic trends in the World and Turkey in terms of age and related indicators and concepts such as life expectancy, fertility rates, and dependency ratios. In the second part, the composition of the elderly population throughout Turkey is examined in terms of socio-economic characteristics and regional/urban/rural/city-scale patterns. In the third part, the housing needs of the elderly are discussed and examples of housing policy and provision from other parts of the world are presented. Then, available housing, housing finance options, and urban facilities and amenities for the elderly in Turkey are reviewed. Lastly, existing public policies and tools for elderly housing in Turkey are explored.

Current Demographic Characteristics and Trends

According to the 2019 Revision of World Population Prospects Report of the United Nations, since 1990, the fastest-growing age group globally is those aged 65 or over among all age groups (0–14, 15–24, 25–64 broad age groups). The share of the population aged 65 years or over increased from 6 per cent in 1990 to 9 per cent in 2019. In 2019, the number of persons aged 65 years or over within the global population was 703 million. This number is projected to double to 1.5 billion in 2050. The share is projected to increase to 16 per cent in 2050 when one in six people will be aged 65 years or over. Globally, the number of people aged 80 years or over, the oldest-old persons, is growing even faster and is projected to triple by 2050, from 143 million in 2019 to 426 million in 2050.

The highest percentage of elderly is in Europe by 25 per cent. However, rapid ageing is observed in all parts of the world, except for Africa. The country with the highest rate of elderly population today is Japan, and it is predicted that Japan will retain that position until the year 2050. As a region, Europe is and is projected to be the oldest until the year 2050. Nevertheless, Asia and Latin America have witnessed an acceleration in the number of elderly in recent decades. Moreover, in terms of the size of the elderly population rather than the rate in overall age groups, Asia is leading the world with 341 million people who are 65 years and older (UN, 2010).

By the year 2050, three-quarters of the world's elderly will be living in developing countries. As women live longer than men and their life expectancy grows faster, the feminization of ageing will be significant as well. The increase in population ageing is faster in urban areas. Between the years 2000 and 2015, the increase was 68 per cent in urban areas, while this rate remained at 25 per cent in rural areas. Approximately 58 per cent of the population aged 60 and over and 63 per cent of the population aged 80 and over live in urban areas as of 2015 (UN, 2015).

Globally, the key drivers of population ageing are declining fertility and increasing longevity. International migration also contributes to changes in population age structures in some countries and regions. However, its effects are mostly projected to be minor. It is projected that between 2015 and 2030, net migration will slow down population ageing by at least 1 per cent in 24 countries or areas and will accelerate population ageing by at least 1 per cent in 14 countries or areas (UN, 2015).

In Turkey, the 65 years and older age group constitute around 9 per cent of the population today. In twenty years, the share is projected to be 16 per cent and in forty years to be 26 per cent, depicting a very rapid increase. The first reason behind population ageing is decreasing fertility rates. Until the 1960s, the number of children per woman in Turkey was around 6.5. This figure has been in decline since then, and in the year 2019, it was 1.88 children per woman (TURKSTAT, 2020a).

The second cause of population ageing is the decrease in death rates, which results in increased longevity. Expected years at birth have been continuously increasing since the 1960s: 42.5 for men and 48.3 for women in 1960; 76.3 years for women, and 71.6 for men in 2010; and 81 for women and 75.6 for men in 2018 (World Bank, 2019).

Migration in Turkey is another factor to consider in population ageing. Inner migration from rural to urban areas in Turkey starting from the early 1950s has resulted in a very rapid urbanization process. While the proportion of rural to urban population was 25 per cent to 75 per cent in 1945, it became 75 per cent urban and 25 per cent rural population by the year 2008. Among others, one important consequence of the migration process on population ageing is the movement of the young population towards the urban and developed regions, leaving more elderly people behind in the rural areas. Another consequence is in more developed regions, migration enabled widespread access to health, education, and security services, which led people to live longer and have increased life expectancy at birth.

Turkey also experiences a considerable flow of international migration. According to TURKSTAT (2020b), the number of immigrants coming to Turkey is nearly 677,000, and 330,000 people emigrated from Turkey in 2019. Most migration recorded is between 25 and 29 years of age (for both emigrant and immigrant categories). It is important to note that according to the "foreign population" definition of TURKSTAT used in the International Migration Statistics, individuals escaping from the war in Syria with a "temporary protection status" are not included. Nevertheless, in Turkey, according to the numbers published by the Directorate General of Migration Management, there were approximately 3.6 million Syrian and almost 400,000 refugees and asylum seekers from other nationalities in 2020 (Ministry of Interior, 2020). Half of the world's refugee population resides in only six countries, and with 14 per cent, Turkey is the leading country (The United Nations Refugee Agency, 2020). It must be noted that there is also a considerable amount of undocumented cases due to the nature of this migration. Therefore, although the UN (2015) Report predicts the effects of international migration to be minor on the pace of population ageing, it is evident that the increasing pace of the flow of migration in Turkey will change the general population structure of the country. Although older persons have a lower share in that migration category, they represent a twice-marginalized group due to additional adaptation problems they face including housing (Demir, 2018). Lewin (2001) points out that besides other challenges, elder immigrants are often unaware of their rights and the alternative forms of housing available to them.

There was at least one person who was 65 years or over in 23.5 per cent of households in Turkey in 2019 (TURKSTAT, 2019a). 1 million 373,000 of these households are the elderly living alone (75.7 per cent are female and 24.3 per cent are male households). The proportion of elderly living alone in the total elderly population has risen from 14.8 in 2009 to 18.2 in 2019. About 33.8 per cent of households include the elderly living alone. The younger aged groups are relatively decreasing in number and the cultural changes indicate that elderly populations, especially women in Turkey, will live alone in the coming decades. For both genders, living alone may lead to isolation, especially with limited socializing options. Particularly for those who do not have social security, adaptation to living alone and to urban living dynamics puts physical, psychological, and socio-economic pressure on the elderly (Köse and Erkan, 2016). Together with the tendency of no child or lower numbers of children, traditional social support mechanisms may fail and thus, loneliness, social exclusion, and long-term care issues should be addressed (Arun, 2014).

According to the results of the 2016 TURKSTAT Family Structure Survey, as far as the preferences on how individuals prefer to live when they are too old to take care of themselves are concerned, 37.6 per cent of the individuals stated that they want to live with their children. The other most preferred option was to receive home care services with 29.4 per cent in the overall population and 27.5 per cent in 65 years and over. Going to a nursing home was chosen by 11 per cent of the overall population. The age group 65 years and over has the lowest proportion in terms of choosing nursing homes with 7.7 per cent. 12.9 per cent of those who were 65 years and over stated that they do not have any idea about how they will live when they get old (Table 13.1).

While the elderly are generally accepted as a vulnerable population, there are often other vulnerabilities accompanying ageing. Disability increases with age. About 8 per cent of men aged between 50 and 64 have at least one restriction of activity of daily living. The figure rises to 17 per cent in those older than 65 (OECD, 2017). For Turkey, the disability rates among the elderly are even more dramatically changing by age, according to the 2011

Table 13.1 Life preferences for elderly ages by gender and age group

	Total	To live in a nursing home	To live with my children	To get home care service	No idea	Other
Turkey total (15+)	100	11	37.6	29.4	21.6	0.4
Gender						
Male	100	10.9	37.7	27.4	23.6	0.4
Female	100	11.2	37.5	31.4	19.6	0.4
Age group						
15–19	100	11.2	30.4	25.5	32.5	0.3
20–24	100	11.6	32.4	26.6	29.1	0.3
25–29	100	11.5	30.8	29	28.4	0.4
30–34	100	12	35.4	29.5	22.7	0.5
35–39	100	11.6	35.1	31.5	21.4	0.4
40–44	100	11.1	38.3	31.4	19	0.2
45–49	100	11.2	38.8	30.7	19	0.3
50–54	100	10.1	41.5	33.4	14.6	0.4
55–59	100	12	39.4	31.5	16.6	0.5
60–64	100	12.7	43.6	28.8	14.5	0.3
65+	100	7.7	51.3	27.5	12.9	0.6

Source: TURKSTAT, Family Structure Survey, 2016 (TURKSTAT, 2016).

data collected via TURKSTAT Population and Housing Census (Figure 13.2). Moreover, older women have a higher proportion compared to the men in the same age group. While the rate of the male disabled population in the age group of 75 and over is 40.9 per cent, this rate is 50.3 per cent in women.

In 2010, those who were 75 years and over had 26 per cent less wealth than the age group between 55 and 64 on average across 19 OECD countries.

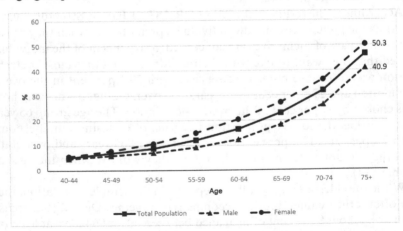

Figure 13.2 Population with at least one disability.

Source: Derived by the Author from Population and Housing Census, 2011 (TURKSTAT, 2011).

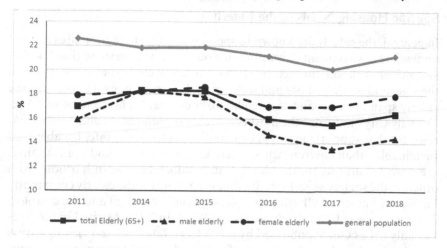

Figure 13.3 Poverty rates among the elderly.

Source: Derived by the Author from TURKSTAT, Elderly Statistics, 2019 (TURKSTAT, 2019a).

There are also inequalities in terms of gender. Old-age poverty risk is higher for older women than men (OECD, 2017). By the year 2018, the poverty rate of the overall population was 21.2 per cent (20.5 per cent for the male population and 21.8 per cent for the female population) in Turkey, and for the elderly, this figure is below average which is 16.4 per cent. Nevertheless, inequality between genders is higher among the elderly than the overall population. While 14.4 per cent of the elderly male population was under the poverty line, the figure was 17.9 per cent for females (Figure 13.3) and the gap between genders increased between 2014 and 2018. The research by Tufan (2006) indicates that 67 per cent of the elderly women in Turkey over age 75 have no income at all, relying solely on the support of their children and relatives. There are further gender differences in ageing. The number of elderly women whose spouses are dead is four times more than the number of elderly men in the same category. While 84 per cent of elderly men are married, only 45 per cent of elderly women are married (TURKSTAT, 2019a).

Overall, the trend of rapid population ageing in Turkey indicates that there will be a different societal composition in the coming decades compared to the younger population structure of the previous decades. There is a lack of experience in dealing with the needs of this new dynamic in all aspects of public policy, including but not limited to housing. Given the presented socio-economic statistics, depicting additional barriers in terms of finances, physical well-being, and social and family networks, elderly people are and will be a vulnerable part of society. Furthermore, there is growing literature in the world on why and how this section of the population will need additional attributes to their housing which mostly cannot be addressed by the existing housing stock.

Specific Housing Needs of the Elderly

In general, the elderly are known to spend more time in their houses and the immediate environment. Therefore, it would be fair to assume that housing and neighbourhood characteristics have a relatively higher importance in their general quality of life and well-being. However, housing needs for the elderly span even beyond the neighbourhood scale. The concept of "Age-Friendly City" indicates the most comprehensive and accessible urban environment that supports active ageing and refers to clean, safe, liveable, and sustainable urban environments, where the environment and opportunities are provided and where the elderly can comfortably live in harmony with others in the society. Age-Friendly City is not only for the elderly community but for the benefit of all urban people, especially children and the disabled. The concept was first introduced by the WHO in 2005. As cited in the Age-Friendly Cities Guide published by the WHO (2007: 9), Age-Friendly City is examined and evaluated within the framework of eight components, which are outdoor spaces and buildings, transportation, housing, social participation, respect, and social inclusion, fulfilling citizenship and participation in the labour force, information and communication, and community support and health services.

The availability of adequate and affordable housing has direct consequences on the quality of life of the elderly. Several studies in the literature emphasize several housing attributes that should be addressed in elderly housing (Hall, 2015; Gallagher, 2015; Luciano et al., 2020). The most significant ones include

- Home accessibility (either during the initial home design or through renovations)
- Home affordability
- Integrated affordable housing and health services
- Transportation, personal care, and health management wellness, socialization, and recreational services and amenities

Moreover, housing conditions should be adaptable to changing needs through the life cycles;

- Age-appropriate home adaptations should be affordable.
- Support mechanisms for mobility should be provided.

The fact that the elderly might need specific qualities in housing leads us to the concepts of "inclusive design", "visitability", and "accessibility". Since part of the elderly population lives with younger household members and some live alone, the housing standards for the elderly should be inclusive to other age groups and also support independent living with increased accessibility and safety options. Accessible housing is broadly defined as having no barriers in terms of architectural design or construction characteristics that prevent ease of access and free movement for people with any kind of limited functions (Alonso, 2002). Enforcing accessibility standards in home construction increases safety as well as inclusivity and thus benefits the elderly and the community at large (Hall, 2015).

Visitability is a standard of home construction, proposing that every home should be visitable by every person in the community, especially those with wheelchairs (Hall, 2015). Since most of the elderly may need supportive equipment in terms of accessibility, visitability standards would directly benefit them and their accompanying or "visiting" family and friends. Three minimum standards of a visitable home are (1) a zero-step entrance, (2) wider doorways on the main floor, and (3) a half bath on the main floor with enough space to fit a wheelchair (Maisel et al., 2008). Other visitability features include grab bars in the first-floor bathroom and lever-style door handles instead of doorknobs (Malloy, 2009). Including these features from the beginning saves money in the future (Maisel et al., 2008).

Visitability ensures minimum requirements to be met; on the other hand, the concept of "universal design" is more comprehensive in that everything within a structure should be designed in a way that is readily accessible to all persons regardless of age, size, or physical ability. This design is concerned with the aesthetic aspect as well, enabling it to be marketable to a broader population (Malloy, 2009). Thus, universal design standards ensure higher accessibility and quality of life not only for the elderly but also accompanying family members and society in general. Standard universal-design features include wider doorways and hallways, lever door handle, slow-closing storm doors, and automatic-opening building doors, wheelchair accessibility, items such as grab bars, tub seats, nonslip or roll-in showers, and ramps, smoke and carbon monoxide detectors, increased lighting, security systems, energy efficiency features such as thicker insulation and specialized double-paned windows, or alternative energy sources such as solar panels (Gallagher, 2015).

The cost of housing is generally thought to be the major factor influencing where the elderly live and their quality of life since unaffordable housing options or the cost of moving might act as barriers for the elderly to move to appropriate housing (WHO, 2007). The elderly are more likely to own their own homes than the general population and to be mortgage-free; however, since most are on fixed incomes, they are "asset-rich and income-poor" (Howden-Chapman et al., 1999).

The longer life span of women means an increase in the number of women of older ages in Turkey and the world in general. Particularly people 75 years and above are mostly women. In their older years, women also have higher rates of disability, higher rates of being single, lower literacy rates, and lower workforce participation. These all indicate that the social, economic, and medical needs and expectations of an elderly woman might differ from the expectations and needs of an elderly man or the general population

(Aksoydan, 2009; Tufan, 2009). Planning and providing housing for the elderly should therefore be diversified according to gender differences, including design and affordability perspectives.

Upgrades and renovations in housing would be another problem area for the elderly. Even regular maintenance of their dwellings would be a challenge for aged populations, and going through major repair work or moving to somewhere else would be a burden for the elderly. According to the results of the TURKSTAT Income and Living Conditions Survey carried out in 2018, a considerable portion of the housing stock in Turkey requires proper necessary upgrades. Around 39.6 per cent of dwellings have heating issues and isolation problems; 36.2 per cent have leaky roofs, damp walls, rotten window frames, and 24.8 per cent experienced air pollution or other environmental pollution problems caused by traffic or industry (TURKSTAT, 2019b).

According to the results of the Household Consumption Expenditures Survey of TURKSTAT for the year 2019, the highest share of consumption expenditures of households has been the housing and rental expenses with 24.1 per cent (TURKSTAT, 2019c). For people living in poverty, this figure is even higher (31.4 per cent). Thus, living in rental properties not only means insecurity of tenure for the elderly but also decreases the chances of making adaptations in the dwellings and the possibility of making savings for homeownership.

The specific housing needs of the elderly might be context-dependent. For instance, in the developing world, most of the elderly live in multi-generational households. Participation in social life and to workforce might also differ in different parts of the world. Healthcare and social security systems also might have implications for housing. While the discussed specific needs of the elderly also exist in Turkey, it should be noted that housing preferences are influenced by cultural differences and family and social relationships.

Current Capacity of Housing Stock in Turkey

Apart from institutional care for the elderly, there is no specific type of housing targeted for the ageing populations in Turkey. Even examples of cooperative housing, which have been initiated by retirees, do not address any specific considerations of ageing. Therefore, the major form of living for the elderly is "ageing in place" without any specific conveniences or measures addressing ageing or moving to nursing homes. "Ageing in place" refers to one's staying in the same housing or living in a housing setting not necessarily designed with ageing populations in mind for the later years of life without moving to assisted living or institutional settings.

In Turkey, Housing Development Administration (TOKİ) is responsible for undertaking mass housing projects for vulnerable populations such as low-income people and disaster victims. Since the year 2015, TOKİ initiated a "Second Spring Housing Program" for the retired citizens' by setting a quota exclusively for the retired. Those who met the income and tenure criteria have been able to apply for the housing developed by TOKİ, which were targeted for low-income families. TOKİ provides housing for retirees in two models (TOKİ, 2016). First, the retirees have a 25 per cent housing quota in 2-bedroom housing projects. Second, residential projects are developed in suitable areas according to the method of collecting preliminary demands.[1] In this model, the demands of retirees are collected, and then the demands are submitted to the governorships and municipalities to request land from them to be transferred to TOKİ to build housing for the retirees.

Conditions to qualify for retirement housing by TOKİ are (TOKİ, 2021):

- To be a Turkish citizen
- To be residing within the provincial/district boundaries of the project for no less than one year or to be registered with the population of the province where the project is located
- Not having bought a dwelling or having used a housing loan from TOKİ before
- Not owning an independent dwelling such as a flat or house (excluding land, vineyard, garden, village house, and workplace), which belongs to either the applicant or spouse and/or children under his/her custody (excluding shared ownership)
- The monthly household income should be at most 7000 Turkish Liras (TRY) for households living in Istanbul and 6500 TRY for all other households as of the year 2021 (this maximum monthly net income includes the income of the applicant, his spouse, and children under their custody, including all kinds of aid they receive for food, transportation)
- The applicant should be retired from the Social Security Institution (in case of death of the pensioner, only the widow of the deceased pensioner can apply)
- If both spouses are retired, only one of them can apply

Since 2015, TOKİ aims to provide approximately 15,000 dwellings for retirees each year (TOKİ, 2018). It should also be noted that since one retired person is considered sufficient for application in each household, the number of retirees who are provided with housing would be higher. Since TOKİ Retiree Housing Program does not require that the age of the retirees be 65 or over and does not collect data based on age, it is not possible to calculate the actual share of those who are 65 years old and over in this housing provision for retirees. By 2019, there were 12 million 200,000 retirees in Turkey (Social Security Institution, 2019). Given the low number of housing provisions, it is evident that the provision is remarkably far from meeting the need. Moreover, these projects of TOKİ address the need for housing for the elderly from the affordability perspective only and lack the quality, design, and lifestyle aspects of ageing. These projects do not have any considerations directly addressing the housing needs of the elderly and thus cannot be considered as "elderly housing", and they do not create age-inclusive housing environments for "ageing in place" in the long term. Moreover, those who retired without social security are excluded from the system, although they need supportive housing policies more.

Since one in every four households has at least one person aged 65 or over, one in every four dwellings should be appropriate for elderly living. There are 38 million 400,000 dwellings in Turkey (TURKSTAT, 2019d). Therefore, ideally, current housing stock should include approximately 9 million 600,000 dwellings suitable for the elderly. Currently, there are no assessment tools or comprehensive studies to evaluate the existing housing stock in terms of meeting the needs and demands of the elderly.

Recently the private sector has also realized the potential of the ageing population, and new projects for the elderly are being undertaken by the construction industry. An earlier example is Ataburgaz Yaşam Kent in Istanbul, designed as an "elderly village", which was designed as a nursing home with a housing community setting. However, the project was not successful due to financial reasons when it was started in 1999 and is now being used as housing for employees (Şener, 2004). A more recent example is the HerDem Project in Silivri, Istanbul (HerDem, 2021). The project is still in the construction phase, with 606 dwellings to be completed by mid-2021, offering rental and ownership options for the elderly (55 years or older) and cleaning, catering, and healthcare packages and recreational, shopping and healthcare facilities. These projects are specifically designed for the elderly and the estates are managed professionally. This kind of housing provision is promising; nevertheless, it serves relatively affluent households and seems to be unaffordable for many other elderly.

In Turkey, the proportion of elderly people living in institutional settings such as nursing homes is only 0.4 per cent (Karakuş, 2018). According to the General Directorate of Services for Persons with Disabilities and the Elderly (EYHGM, 2020) data, the total number of public and private nursing homes in Turkey is 428, with 27.575 elderly under care. Approximately 36 per cent of the total nursing homes are affiliated with the state (The Ministry of Family and Social Services[2]) and 58 per cent are run by the private sector. 51 per cent of the elderly live in public nursing homes, while 39 per cent are in private nursing homes. The number of people under care by 2020 is 27.117, which indicates a more than double increase since 2002. Due to the increasing demand for elderly care and demand for different physical settings, new service models for the elderly have also been implemented in the public nursing homes by the Ministry, offering more home-like settings such as the Small Houses Project, Housing Estate Project, Courtyard Nursing homes, Elderly Living Homes, and Elderly Support Program. The elderly can stay alone or with their spouses in Elderly Living Houses, which are detached houses or apartments, generally in city centres and larger communities, where 3–6 people stay together. The courtyard model is designed for elderly people who can carry out their daily work on their own and who need more physical activity. The Housing Estate type project is a nursing home model that is designed to serve the elderly in need of care due to weakened bodily functions.

Available Tools and Potentials of the Housing Policy in Turkey

The first international document in the world, which includes provisions on the assurance of elderly rights, is the "United Nations Declaration of Human Rights" issued in 1948. This declaration includes regulations regarding the individual's right to an appropriate standard of living, social security, and social services in old age. Article 25 of the Universal Declaration of Human Rights states that the elderly have the right to ageing safely in healthy living environments. In the first World Assembly on Ageing, held in Vienna in 1982, countries adopted the International Plan of Action.

The housing need of the elderly has recently been addressed by the public policy in Turkey. In terms of developing policies on the housing of the elderly, one of the key documents in Turkey is the Situation of Elderly People in Turkey and the National Action Plan on Aging prepared by the State Planning Organization in 2007. The studies in Turkey were initiated following the "2002 Madrid 2nd International Plan of Action on Ageing", which was prepared by the United Nations, to improve the quality of life of the elderly population, ensure their social integration, resolve livelihood and health problems, and create policies covering all age groups. Besides referring to several other aspects including healthcare, social inclusion, and employment, Turkey's National Action Plan on Aging (State Planning Organization, 2007) emphasizes creating Elderly Friendly Cities by designing and producing age-inclusive urban furniture and improving the ergonomics of vehicles for the elderly and the disabled.

In 2011, EYHGM was established to guide policies and services for the elderly. The first Five Year Development Plan to mention the elderly with regards to their needs for housing and urban areas is the Tenth Five Year Development Plan (Ministry of Development, 2013). As part of the Development Plan, the Ministry of Development[3] prepared a Special Commission Report on Ageing in 2014. This study emphasizes nursing homes as the mainstream solution for the elderly, yet draws attention to actions that need to be taken in the housing sector including safety and security measures by ensuring physical standards for new housing and landscape such as elevators, light buttons, heating devices with sensors, and non-slippery floors. The report also suggests TOKİ develop housing specifically targeting the elderly.

The second Special Commission Report on Ageing as part of the Eleventh Five Year Development Plan in the year 2018 employs the 'Active Aging'" concept as a framework. This concept highlights an inclusive approach adopted by the WHO towards the end of the 1990s and emphasized in the 2002 Madrid International Plan of Action on Ageing.

The report suggests municipalities make physical adjustments in existing buildings to prevent home accidents and facilitate the life of the elderly. This report also uses the term age-friendly cities and suggests that public spaces should be suitable for the elderly to achieve active ageing (Ministry of Development, 2018a).

Other Special Commission Reports, such as the report on Housing Policies (Ministry of Development, 2018b), state that social housing projects will be supported to meet the housing needs of disadvantaged groups such as low-income, disabled, elderly, women, youth, and homeless. Adaptable housing and universal housing approaches should be adopted. Physical measures and standards should be developed, city guides should be prepared for the disabled and elderly.

The Special Commission Report on Quality of Urban Life (Ministry of Development, 2018c) highlights that the participation of the increasing elderly population in urban life should be supported. Access of the elderly to all areas of the city should be encouraged; participation in social, cultural, and artistic activities should be supported. Elderly care centres should not be built in places far from the city and urban life so that the elderly can keep their daily life habits.

The First "Council on Ageing" was organized in 2019, and the year 2019 was declared as "the year of older persons" in Turkey. One of the commissions of the council is Age-Friendly Cities and Local Governments. According to the commission's report (2019), local governments should provide logistics, projects, and financial support to the citizens, especially in the process of reorganizing old and inappropriate houses in a way to facilitate the lives of the elderly. For the elderly living in poverty, prefabricated houses should be built in the places determined by the municipalities. In urban transformation projects, the displacement of the elderly should be prevented. They should be considered as a separate target group and their rights should be protected. Some of the communities redeveloped by urban transformation projects should be designed considering age-friendly principles.

Local governments today offer a variety of services to the elderly ranging from daily care, house cleaning, minor repairs, day centres offering educational or cultural programs, cultural activities, and such in addition to institutional care. Some of the elderly are supported financially. Nevertheless, these programs are not standardized, vary from one local government to another, are prone to change after political elections, and rarely involve any form of direct housing provision including rentals, rental assistance, or affordable purchasing options. Local governments have not yet approached elderly housing as a housing policy problem.

In the housing sector, there are some other financial advantages and services provided for the elderly in Turkey. Those who do not have any income other than their pension from social security institutions are exempted from paying real estate tax provided that they own only one dwelling, which is less than 200 m^2. Those aged 65 and over benefit from free urban public transportation as well, which removes the financial barrier of accessibility to services that are not provided in their immediate housing environment, although there are no specific arrangements to remove accessibility barriers.

Housing and Housing Policy Examples from the World

Examples of elderly housing throughout the world reveal that apart from institutional care options for the elderly, housing options can be grouped into two major categories: independent living environments (mostly for younger old) and assisted living (mostly for oldest-old). Different housing strategies for the ageing population in independent living environments in many countries include active adult communities, senior apartments, or retirement communities. Assisted living options, on the other hand, offer security measures, healthcare, and nutrition services, among other forms of assistance. Institutional care options include nursing homes and elderly hospitals. The terms used for different types of housing options can sometimes vary between countries and regions.

Many cities and countries develop different solutions for the housing problem of the elderly. These include the direct provision of housing, rental assistance, transportation, housing maintenance, and many other programs. Many of the strategies address the physical barriers of the elderly and financial hardships. Affordability of housing includes the cost of renting or purchasing a home, plus repair, maintenance, and adaptations with the aim of security, safety, comfort, and functionality. Thus, while affluent older households have many options, many others need supportive public policies. Other innovative tools such as the reverse mortgage—where the elderly borrow against the value of their homes—are also used to improve the financial status of the elderly who are homeowners. This strategy is widely used in Europe, the United States, and Australia (Ong, 2008; Moscarola et al., 2015; Knaack et al., 2020).

Among all options, ageing in place is the most common way of housing for the elderly. Ageing in place enables staying at home as people age and has the advantage of keeping the elderly in a familiar place where they know their neighbours and the community. Most OECD countries are committed to reducing the number of people living in institutions (OECD, 2003). Nonetheless, to maintain sufficient quality of life, all housing environments should be designed with all ages inclusive employing universal design approaches. The elderly can take advantage of home care services including household maintenance, transportation, home modifications, personal care, health care, and socializing day programs. Ageing in place requires a functional mix of roles and responsibilities in decision-making systems. In Australia, Japan, Denmark, and Spain, central governments are responsible for defining overall policy guidelines to be implemented at local levels. In Finland, the UK, and New Zealand, local governments define and manage social and housing policies with possible central government financial assistance (OECD, 2003). These experiences would serve as important examples for Turkey to develop housing policies for the elderly. Since previous experiences emphasize the importance of policies in ageing in place and since housing cannot be considered in isolation, age-friendly living environments would require appropriate policies in transportation, physical, and social infrastructure as well.

Results and Discussion

This study was designed to question the potentials and limitations of the housing stock and housing policy in Turkey to cope with the implications of the ageing population. The study first analyzed the specific housing needs of the elderly. The results of the study revealed that the demographic characteristics and housing needs of the elderly vary from the general population in several ways. In older age groups, disabilities increase, health conditions deteriorate, and incomes are mostly fixed and lower. There are more prominent gender differences, and the number of elderly women is higher compared to that of elderly men. Living together or close to family and friends is mostly a priority for the elderly; thus, results indicate that housing for older populations should also be appealing to other age groups. Moreover, in different phases of old age, the elderly should also have options including moving to assisted and institutional settings.

The study also questioned the current capacity of housing stock in Turkey. According to the results of the study, the current capacity of the housing stock is hardly a match to the needs of the elderly. Public support programs of housing address only the "retirees" rather than the "elderly", meaning that they are not inclusive of those who are not retired from a social security system. Furthermore, they lack adequate physical design considerations for the elderly, and the number of produced dwellings is disproportionately less than the population of the elderly. Housing projects developed by the private sector for the elderly are emerging very recently as a new form of housing. Yet, above other concerns, they might only be a solution for the more affluent elderly populations. In terms of "ageing in place", inclusive and universal design principles are hardly part of most neighbourhoods and cities; thus, they do not provide healthy and high-quality environments for the elderly.

Lastly, the study examined the available tools and potentials of the housing policy in Turkey. In terms of national and local policies, there is a recent increased interest in ageing as a result of demographic trends and the impact of the international agenda in terms of ageing policies. Yet, it is generally considered as healthcare, day-care, or a social security problem. Housing policy for the elderly is only referred to in vague terms. The elderly are mostly handled within the disabled category or are cited as the vulnerable segment of the society together with children, women, and people living in poverty. The needs of the elderly are only recognized on the level of need for basic needs and care. Their lifestyle choices, life satisfaction, and creating active and healthy living environments are ignored. The first initiatives at the national level suggest that this increased awareness will manifest itself in further studies and actions including more clearly defined principles, housing and urban design guides, and implementations in housing provision for the elderly.

The results of the study indicate a possible housing crisis for the elderly population in Turkey in terms of number, affordability, and quality. The results of the analysis also revealed that housing solutions for the elderly require certain issues and groups to be considered specifically. The first is the elderly women. Gender is a major consideration since the share of women is higher in older ages, and also, women have higher rates of poverty, disability, and solitary life. This would imply a need for supplementary policies for women and consider their specific needs and lifestyle in the design and location choices of housing. Second, the elderly population living in rural areas is critical. Lower population density and more geographically dispersed populations make it more difficult and expensive to create and maintain a comprehensive service infrastructure in rural areas. Elderly care services provided by the state and private sector in urban areas are completely provided by families in rural areas, meaning a greater risk of social isolation, reduced mobility, lack of support, and health care deficits. The elderly who live in rural areas and who have lost the young population around them due to migration constitute a potential risk group that will be in serious need of care. The third group is the elderly living alone, who have a significant share among people living alone. This would imply smaller housing designs that target the elderly as the major consumers. With health assistance needs and social isolation problems in mind, housing should be supported by additional support services including healthcare, access to transportation, and social infrastructure.

Today, one in every four households has at least one elderly person. Moreover, younger household members are also candidates for being old in the coming decades. This implies the need for policy and strategies for developing new housing designated primarily for the elderly and also for re-evaluating the existing housing stock and neighbourhoods for "ageing in place". Studies reveal that ageing in place strengthens social relationships and participation in society and is more inexpensive, efficient, and beneficial for physical and mental health. Thus, all housing and neighbourhoods should be analyzed, designed, and/or improved with age-friendly principles. Major components of this analysis should involve the current suitability of housing, future suitability of housing, maintenance, renovation, and the perceived housing quality.

The demographic trend of the ageing population in the world will result in important long-term societal changes, and Turkey needs to address this trend rapidly to prevent a possible housing crisis for the elderly. However, housing policies so far have not adequately addressed this emerging but striking housing need. Turkey needs an urgent and comprehensive roadmap with a strong housing policy. Yet, the housing construction sector is very dynamic in Turkey. This includes not only the construction of new housing but also several urban redevelopment and transformation projects. This would be an immense opportunity for adopting and implementing standards for elderly friendly housing environments and would mean being prepared for the elderly populated cities in the years to come. Housing provision should be handled together with management, finance, and investment dimensions and should also be supported by policies on the financial, physical, social, and mental well-being aspects of ageing. This study served to provide a foundation and a framework to raise awareness about the housing needs of the elderly as a separate housing policy area. Further case studies should be performed to gain a better insight into ageing populations and their housing needs.

Notes

1 TOKİ uses this method of collecting preliminary demands in order to prevent the idleness of investments to be made in projects towards settlements with a population below 40,000 (TOKİ, 2016).
2 Previously Ministry of Family, Labour and Social Services became The Ministry of Family and Social Services in the year 2021.
3 The Ministry of Development and Budget Directorate of Ministry of Finance were merged into the Presidency of Strategy and Budget in the year 2018.

References

Abramsson M, Andersson E (2016) Changing preferences with ageing—housing choices and housing plans of older people. *Housing, Theory and Society* 33(2): 217–241.

Aksoydan E (2009) Are developing countries ready for ageing populations? An examination on the socio-demographic, economic and health status of elderly in Turkey. *Turkish Journal of Geriatrics* 12(2): 102–109.

Alonso F (2002) The benefits of building barrier-free: A contingent valuation of accessibility as an attribute of housing. *European Journal of Housing Policy* 2(1): 25–44.

Arun Ö (2014) 2050'ye Doğru Yaşlanan Türkiye'yi Bekleyen Riskler. *Selcuk University Journal of Institute of Social Sciences* 32: 1–12.

Council on Ageing (2019) Commission Reports of First Council of Aging. Available at: https://ailevecalisma.gov.tr/media/40504/i-yas-lilik-s-urasi-07022020-pdf-pdf.pdf. Accessed 15 May 2021.

Demir SA (2018) Sakarya'daki Yaşlı Suriyelilerin Göç Deneyimleri ve Sorunları (Migration Experiences and Problems of Older Syrians in Sakarya). *Göç Dergisi* 5(2): 205–218.

European Commission (2019) Aging Europe: Looking at the Lives of Older People in the EU. Publications Office of the European Union. Available at: https://ec.europa.eu/eurostat/documents/3217494/10166544/KS-02-19%E2%80%91681-EN-N.pdf/c701972f-6b4e-b432-57d2-91898ca94893. Accessed 15 May 2021.

EYHGM (2020) Engelli ve Yasli Istatistik Bulteni, Mayis 2020. Available at: https://www.ailevecalis ma.gov.tr/media/51832/mayis-istatistik-bulteni.pdf. Accessed 15 May 2021.

Gallagher J (2015) The evolution of elder housing design and development. *Maine Policy Review* 24(2): 42–48. Available at: https://digitalcommons.library.umaine.edu/mpr/vol24/iss2/9. Accessed 15 May 2021.

Hall T (2015) Inclusive design and elder housing solutions for the future. *NAELA Journal* 11(1):61–71.

Herbers DJ, Mulder CH (2017) Housing and subjective well-being of older adults in Europe. *Journal of Housing and the Built Environment* 32(3): 533–558.

HerDem, (2021) HerDem Corporate Website. Available at: https://www.herdemsyk.com. Accessed 15 May 2021.

Houben PPJ (2001) Changing housing for elderly people and co-ordination issues in Europe. *Housing Studies* 16(5): 651–673.

Howden-Chapman P, Signal L and Crane J (1999) Housing and health in older people: Ageing in place. *Social Policy Journal of New Zealand 13*: 14–30.

Karakuş B (2018) *Türkiye'de yaşlılara yönelik hizmetler, kurumsal yaşlı bakımı ve kurumsal yaşlı bakımında illerin durumu.* Ankara: Aile ve Sosyal Politikalar Bakanlığı Engelli ve Yaşlı Hizmetleri Genel Müdürlüğü.

Knaack P, Miller M and Stewart F (2020) *Reverse Mortgages, Financial Inclusion, and Economic Development: Potential Benefit and Risks.* The World Bank. Available at SSRN: https://ssrn.com/abstract=3528858. Accessed 20 May 2021.

Köse N, Erkan NÇ (2016) Kentsel Mekân Örgütlenmesinin Yaşlıların Kentsel Etkinlikleri Üzerindeki Etkisi, İstanbul ve Viyana Örneği. *METU Journal of the Faculty of Architecture* 31(1): 39–66

Lewin FA (2001) The meaning of home among elderly immigrants: Directions for future research and theoretical development. *Housing Studies* 16(3): 353–370.

Luciano A, Pascale F, Polverino F et al. (2020) Measuring age-friendly housing: A framework. *Sustainability* 12(3): 848.

Maisel J, Smith E and Steinfeld E (2008) Increasing home access: Designing for visitability. *AARP Public Policy Institute* 14: 1–34.

Malloy RP (2009) Inclusion by design: Accessible housing and mobility impairment. *Hastings Law Journal* 60: 699.

Ministry of Development (2013) Tenth Development Plan Report. Available at: http://www.sbb.gov.tr/wp-content/uploads/2018/11/Onuncu-Kalk%C4%B1nma-Plan%C4%B1-2014-2018.pdf. Accessed 15 May 2021.

Ministry of Development (2018a) Special Commission Report on Ageing. Available at: https://www.sb b.gov.tr/wp-content/uploads/2020/04/YaslanmaOzelIhtisas KomisyonuRaporu.pdf. Accessed 15 May 2021.

Ministry of Development (2018b) Special Commission Report on Housing Policies. Available at: https://www.sbb.gov.tr/wp-content/uploads/2020/04/Konut PolitikalariOzelIhtisasKomisyonuRaporu. pdf. Accessed 15 May 2021.

Ministry of Development (2018c) Special Commission Report on Quality of Urban Life. Available at: https://www.sbb.gov.tr/wp-content/uploads/2020/04/KentselYasam KalitesiOzelIhtisasKomisyonuRap oru.pdf. Accessed 15 May 2021.

Ministry of Interior (2020) Temporary Protection Statistics. Directorate General of Migration Management. Available at: https://en.goc.gov.tr/temporary-protection27. Accessed 15 May 2020.

Moscarola FC, D'Addio AC and Fornero E et al. (2015) Reverse mortgage: A tool to reduce old age poverty without sacrificing social inclusion. *Ageing in Europe-Supporting Policies for an Inclusive Society* 235–244.

OECD (2003) *Ageing, Housing and Urban Development*. Paris: OECD Publishing Available at: https://doi.org/10.1787/9789264176102-en. Accessed 15 May 2021.

OECD (2017) *Preventing Ageing Unequally*. Paris: OECD Publishing. Available at: http://dx.doi. org/10.1787/9789264279087-en. Accessed 15 May 2021.

Ong R (2008) Unlocking housing equity through reverse mortgages: The case of elderly homeowners in Australia. *European Journal of Housing Policy* 8(1): 61–79.

Social Security Institution (2019). Annual Statistics. Available at: http://www.sgk.gov.tr/wps/portal/sgk/tr/kurumsal/istatistik/sgk_istatistik_yilliklari. Accessed 15 May 2021.

Sixsmith J, Sixsmith A and Fänge AM et al. (2014). Healthy ageing and home: The perspectives of very old people in five European countries. *Social Science & Medicine* 106: 1–9.

State Planning Organization (2007) The Situation of Elderly People in Turkey and National Action Plan on Ageing. Available at: http://ekutup.dpt.gov.tr/nufus/yaslilik/eylempla-i.pdf. Accessed 22 September 2020.

Şener B (2004) *Türkiye'de Yaşlı Konutlarının Yapılabilirliğine Yönelik Fizibilite çalışması*. Unpublished Master's Thesis. ITU Graduate School of Natural and Applied Sciences.

The United Nations Refugee Agency (2020) Refugee Data Finder. Available at: https://www.unhcr.org/refugee-statistics/. Accessed 15 May 2021.

TOKİ (2016) Corporate Profile. Available at: http://i.toki.gov.tr/content/images/main-page-slider/16012017212815-pdf.pdf. Accessed 15 May 2021.

TOKİ (2018) TOKI News. Available at: https://www.toki.gov.tr/haber/toki-baskani-turan-bu-yil-15-bin-emekli-vatandasimizi-ev-sahibi-yapacagiz. Accessed 15 May 2020.

TOKİ (2021) Requirements for Low-Income Retirement Housing. Available at https://www.toki.gov.tr/basvuru-sartlari. Accessed 15 May 2021.

Tufan İ (2006) *Yaşlılıkta Bakıma Muhtaçlık ve Yeni Bir Bakım Kültürü* (Need for Care and a New Culture of Care in Old Age). Antalya: Gero Yayınları.

Tufan İ (2009) Ageing and the elderly in Turkey—results of the first age report from Turkey. *Zeitschrift Fur Gerontologie und Geriatrie* 42(1): 47–52.

TURKSTAT (2011) Population and Housing Census, 2011. Available at: https://www.ailevecalisma.gov.tr/media/5677/nufus-ve-konut-arastirmasi-engellilik-arastirma-sonuclari.pdf. Accessed 10 May 2021.

TURKSTAT (2016) Family Structure Survey, 2016. Available at: https://tuikweb.tuik.gov.tr/PreHaberBultenleri.do?id=21869. Accessed 15 May 2021.

TURKSTAT (2019a) Elderly Statistics 2019 News Bulletin, No 33712. Available at: https://data.tuik.gov.tr/Bulten/Index?p=Istatistiklerle-Yaslilar-2019-33712#. Accessed 15 May 2021.

TURKSTAT (2019b) Income and Living Conditions Survey 2018, News Bulletin, No 30755. Available at: https://data.tuik.gov.tr/Bulten/Index?p=Gelir-ve-Yasam-Kosullari-Arastirmasi-2018-30755. Accessed 15 May 2021.

TURKSTAT (2019c) Household Consumption Expenditure Survey. News Bulletin, No 33593. Available at: https://data.tuik.gov.tr/Bulten/Index?p=Hanehalki-Tuketim-Harcamasi-2019-33593. Accessed 15 May 2021.

TURKSTAT (2019d) Address-Based Population Registration System Results. Available at: https://tuikweb.tuik.gov.tr/PreTablo.do?alt_id=1059. Accessed 15 May 2021.

TURKSTAT (2020a) Birth Statistics 2019 News Bulletin, No 33706. Available at: https://tuikweb.tuik.gov.tr/PreHaberBultenleri.do?id=33706. Accessed 15 May 2021.

TURKSTAT (2020b) International Migration Statistics News Bulletin No 33709. Available at: https://tuikweb.tuik.gov.tr/PreHaberBultenleri.do?id=33709. Accessed 12 August 2020.

United Nations (2002) Madrid International Plan of Action on Ageing. Available at: https://www.un.org/esa/socdev/documents/ageing/MIPAA/political-declaration-en. pdf. Accessed 15 May 2021.

United Nations (2010) *World Population Ageing-2009*. New York: Department of Economic and Social Affairs, Population Division.

United Nations (2015) *World Population Ageing 2015 (ST/ESA/SER.A/390)*. New York: Department of Economic and Social Affairs, Population Division.

United Nations (2019) *World Population Prospects: The 2019 Revision Highlights*. New York: Department of Economic and Social Affairs, Population Division. Available at: https://population.un.org/wpp/Publications/Files/WPP2019_10KeyFindings.pdf.

World Bank (2019) Indicators: Life Expectancy at Birth, Turkey. Available at: https://data.worldbank.org/indicator/SP.DYN.LE00.IN?locations=TR. Accessed 15 May 2021.

World Health Organization (2007) *Global Age-Friendly Cities: A Guide*. Geneva: World Health Organization.

World Health Organization (2011) Global Health and Aging. Available at: https://www.who.int/ageing/publications/global_health.pdf?ua=1. Accessed 10 May 2021.

World Health Organization (2015) *World Report on Ageing and Health*. World Health Organization. Available at: http://apps.who.int/iris/bitstream/handle/10665/186463/9789240694811_eng.pdf?sequence=1. Accessed 15 May 2021.

14 Housing and Transport Cost Burden

The Experience of Tenant Households in Turkey

Esma Aksoy Khurami and Ö. Burcu Özdemir Sarı

Introduction

The formation and regulation of housing markets define the trajectories of the modes of tenure. In Turkey, the tendency was towards the glorification of homeownership. Policies targeted an increase in the share of owner-occupier households by means of new housing provision, the transformation of existing housing stock, housing sale programmes, and taxes and savings incentives (Aksoy Khurami, 2021). The provision of rental units is expected by private households owning single and multiple housing units. In other words, policies to create social, affordable, and public rental housing stocks are not on the agenda of governments. Except for a few examples, no wide range of regulations is imposed on the private rented sector, and tenant households and rental units are left to their fates. While the share of tenant households remained almost constant, with 22.3% in 2004 and 23.4% in 2018, the percentage in owner-occupier households decreased from 71.1% in 2004 and 60.7% in 2018 in defiance of homeownership-promoting policies (TURKSTAT, 2004, 2018). Although it corresponds to a quarter of total households, due to the limits regarding countrywide data and the dynamic structure of the private rented housing market, studies focussing on rental units and tenants have remained few.

The problems in the private rented sector in Turkey are rarely approached in terms of the inadequate supply of units, the quality of private rental units, and the legal rights and responsibilities of tenants and landlords (Sarıoğlu-Erdoğdu, 2014, 2015; Alkan and Uğurlar, 2015; Uğurlar and Özelçi Eceral, 2017). In line with the growing international discussions on the increase in housing and transport expenditures of households, studies conducted in Turkey addressed the affordability of different modes of tenure, especially owner-occupier households (Aşıcı, Yılmaz, and Hepşen, 2011; Coşkun, 2020). They revealed that homeowner households and privately rented unit tenants are experiencing a heavy burden of housing and transportation expenses (Özdemir Sarı and Aksoy Khurami, 2018).

Unlike previous examples, the present study attempts to reveal items one by one affecting the burden of housing and transport expenditures of tenant

DOI: 10.4324/9781003173670-18

households in Turkey. It employs Household Budget Survey-2018 cross-sectional data produced by the Turkish Statistical Institute and examines transport and housing expenditure items. In that respect, housing expenditures items are evaluated in three categories: (i) housing rent, (ii) bills (water, electricity, and heating), and (iii) essential maintenance, subscription, and service costs. Transport expenditures are categorized as (i) public transport cost (passenger transport by train, underground, tram, bus, flights, sea vehicles, inland waterway, compound passenger transportation, cable railway, cable car, and chairlift) and (ii) private transport cost (first- and second-hand motorized and non-motorized vehicles, their tyres, spare parts and accessories, fuels, oils, maintenance and repair costs, rental and usage cost of garages, parking spaces and personal vehicles and taxi costs, tolls and parking meters, and transport and shipping services).

The findings show an increase in the share of tenant households experiencing the burden of housing and transport expenditures from lower to higher-income groups. The shares of housing rent and private transport expenditures in tenant households' budget are revealed as the main differences among households subjectively indicating they have a burden and those indicating they do not. Finally, the mismatch between the number of bedrooms in the housing unit and household size is not found to be an identifier of the burden of housing and transport expenditures but of the duration of occupancy. The greater duration tenant households reside, the less housing rent and burden are observed.

This chapter consists of five sections. Following the introduction, section 2 focusses on the approaches to the measurement of housing and transport expenditures of households. Then observations on the private rented housing sector in Turkey with experimental and legal aspects are covered in section 3. Section 4 focusses on the investigation of housing and transport expenditures of tenant households. The last section discusses the current situation regarding the burden in Turkey and concludes with the findings and drawbacks of the chapter.

Housing and Transport Expenditures

The cost of housing, also referred to as the burden of housing expenditures and housing affordability, is among the main focusses in urban studies in the 21st century due to the cumulative effects of the financial and housing crises. At first glance, these problems are defined and approached based on the distinctive assumptions and perceptions of households. To evaluate the burden of housing expenditures on households, a housing unit is first expected to meet normative local or accepted standards (Stone, 2006). Based on the determined standards, Edgar et al. (2002: 59) point out that "…. a simple definition of affordability is that the price of that dwelling unit (net of state subsidy) is not an unreasonable burden on households' income". In that respect, the burden of housing expenditures is partly associated with the ongoing cost of housing and the income levels of households (Leishman and Rowley, 2012: 379).

The methods used to measure the level of the burden of housing expenditures focus on the different treatments of income and housing and sometimes transport expenditures. Several methods have been developed, but there is "no clear-cut answer" (Galster and Lee, 2021: 12) as to how to measure the burden of housing costs, and it is all well documented in the literature (Bogdon and Can, 1997; Jewkes and Delgadillo, 2010; Sendi, 2014; Aksoy, 2017; Haffner and Hulse, 2021). Some studies focussed on the household level to show the spatial and characteristic distribution of households living in affordable and unaffordable housing units (Yates, 2006; Collinson, 2011; Krapf and Wagner, 2020). Some aggregate data to reveal the country-/citywide results of housing policies (Quigley and Raphael, 2004; Beer, Kearins, and Pieters, 2007; Preece, Hickman, and Pattison, 2020).

Commonly, the definition of the burden of housing cost, in any case, is based on a household spending more than 30% of their budget on housing expenditures (Belsky, Goodman, and Drew, 2005), the ratio approach (Thalmann, 1999; Bramley, Munro, and Pawson, 2004). Considering only the share of housing and transport expenditures in the household budget is regarded as superficial when households' burden and housing conditions are evaluated. Stone (1993, 2006) stated that using the ratio alone will be insufficient to explain the diversity of household expenditure and living conditions. For example, although the ratio remains above the proportionally determined limit, a high-income household's residual income after housing expenditures may be sufficient to allow a decent quality of life. On the other hand, even if the ratio is below the specified threshold for a low-income household, the remaining income may be insufficient to provide the same quality of life. Similarly, Padley, Marshall, and Valadez-Martinez (2019) underlined the thorniness of the subjective evaluation of housing affordability due to the different perceptions of households between the need and preference of housing. They also argue that the availability of several housing options at different price and rent levels affects this evaluation.

Following the ratio approach, the residual income approach, a new dimension in the measurement of housing affordability, embraces the rationale of being able to meet non-housing needs with residual income after the deduction of housing expenditure (Bourassa, 1996; Malpass and Murie, 1999). It is frequently used to assess the burden of housing-related expenditures on households' decent quality of life. For example, Revington and Townsend (2016) applied the residual income approach to measuring rental market affordability in Vancouver and Montreal for couples with and without children. They found that staying inside and outside of rapid transit walking areas was decisive in terms of meeting non-housing expenditures.

The burden of housing expenditures is associated with households that intend to be homeowners due to the effects of the financial crisis rather than other modes of tenure, especially tenant households (Dewilde, 2018;

Byrne, 2019) because economic crises affect household incomes through the loss of regular jobs and downstream income. In addition, they have a severe impact on the housing markets, such as causing an enormous increase in housing prices, changing availability of mortgage credits, and ultimately foreclosures. However, the loss of regular jobs and downstream income has an immediate effect, as observed through the purchasing of foreclosure housing units by prominent estate agent firms: a decrease in the number of rentable housing units with low and no profit (Fields, 2018). Due to the spatial expansion in the cities and the high rents of the housing units located in desirable and accessible locations, households were forced to choose between different options: (i) pay higher rents but have easy access, (ii) pay less for housing rent but spend more time and money to commute. That is why in addition to the housing rent, transportation expenditures and other housing costs are considered the main factors behind the affordability crisis of tenant households (Bogdon and Can, 1997; Hulchanski, 1995; Yates and Gabriel, 2006; Joyce, Mitchell, and Norris Keiller, 2017; Singer, 2020). Leone and Carroll (2010) mentioned the rent level, mainly since the 1970s, but every policy discouraging rental housing provision, directly or indirectly, has supported other tenure modes. As a result, a decrease in rental stock and the problem of rental and affordable housing emerged. Along with the burden of housing expenditures, transport costs are inspected to study the affordability of housing units having various locational advantages and disadvantages. They are considered to estimate the role of intra-urban rail services and auto-oriented neighbourhoods (Singer, 2020), automobile ownership (Walks, 2018), and fare structure (Nassi and da Costa, 2012) in the affordability of low-income level households. Dodson et al. (2020) overviewed the effect of affordable rental housing provision on the commuting burden of tenant households in Melbourne and Sydney. Although there has been an increase in the number of studies, the need for more is obvious; their findings align with the previous evidence, which shows the spatial mismatch between where low-income households can afford and where they are employed.

Unlike the recently emerged developed country examples of housing affordability studies, housing affordability is associated chiefly with owner-occupier concerns due to the dominance of homeownership-promoting housing policies in Turkey. Aşıcı, Yılmaz, and Hepşen (2011) and Coşkun (2020) evaluated homeownership affordability of households from the macro perspectives of economics. Aşıcı, Yılmaz, and Hepşen (2011) are among the first carrying out a housing affordability study in Turkey. They emphasized housing sale prices, loan-to-value ratio, household income, mortgage interest rate, and maturity of the mortgage payment for seven metropolitan cities in Turkey by utilizing many secondary data sources for the two years between 2007 and 2009. However, using the loan-to-value ratio in housing affordability index measurement can be considered to result in an overestimation of the low share of households benefiting from mortgage credit in Turkey. Coşkun (2020) recently focussed on affordability in Turkey and some metropolitan cities only from a homeownership perspective through creating different indexes and ratios (employing housing cost, median housing price, income, and loan-to-value ratios). The findings pointed to the unaffordability of homeownership except for the highest-income households; yet the study did not utilize household-level data and generalized the measurement variables for all households. Emekci (2018) applied a life-cycle analysis of housing affordability to sustain low-cost housing for low-income households in Turkey. The present study is conducted at the architectural scale of dwelling and housing units. As a result of many assumptions and variations of decisions, optimal solution alternatives have been developed.

A group of studies was conducted for countrywide analysis of housing affordability (Özdemir Sarı and Aksoy, 2016; Aksoy, 2017; Özdemir Sarı and Aksoy Khurami, 2018). They employed the Survey of Income and Living Conditions and Household Budget Survey to measure the burden of housing and transport expenditures for different income groups, modes of tenure, and NUTS (Nomenclature of Territorial Units for Statistics) regions.

A Missing Piece of the Housing Puzzle: The Private Rented Sector in Turkey

There is no constant and regular private rented housing stock in Turkey. Dwelling units are leased by their owners for rental income and profit (Balamir, 1999). When an owner decides to sell or not to rent a housing unit, there is no obligation to ensure the continuity of the rental housing stock and vice versa. Yet, the private rented sector is large enough to accommodate almost a quarter of urban households in the current situation. This situation can also be interpreted as the rental stock not being created according to any need or demand. Above all, the spatial distribution of rental units depends on the owner's decisions. Hence, units are not provided in some locations and there is an excess provision in others. This directly influences the transport cost of households; more importantly, scarcity is decisive in the housing rent and so housing expenditures.

Tenancy in Turkey is not directly defined by any specific law right now. However, some principles from the Real Estate Rentals Law (Law no. 6570) repealed in 2011 are included in the Flat Ownership Law (Law no. 634) and Law of Obligations (Law no. 6098). The leasing agreement also includes a brief articled summary of the Flat Ownership Law and Law of Obligations defining the owner's and the tenant's relationship and responsibilities. Some improvements are made to protect tenant households from owner's arbitrary decisions; yet many points defining the rights and responsibilities of tenant households are ignored. For example, the Law of Obligations sets the amount of deposit payment as the cost of 3 months of housing rent. However, in daily experience, some owners prepare a promissory note for more than this amount to meet the cost of repairs in the event of any damage. It constitutes an extra burden of possible housing expenditures for tenant households.

The regulations and policies implemented for private rented housing sectors have been limited and overshadowed by those for homeownership promotion. Tenant households are at the mercy of their landlords and basic regulations including the annual increase in rent level. The leasing agreement is the only document indicating the housing unit's current condition (the material and quality of frames and doors, the condition of the painted walls, etc.) and fixtures (the brand and condition of the central heating boiler and shower cubicle, etc.) in written statements. Although it is compulsory, many owners do not want to have a leasing agreement or official payments through banks to avoid paying tax on their rental income. Moreover, the lack of a leasing agreement can end in the forced eviction of tenant households upon claims by the owner. The conditions of housing and fixtures are not usually photographed in a leasing agreement; therefore, a tenant can be exposed to the arrogation of an owner about damage to the house and its fixtures. This influences the level of housing expenditures for tenant households.

The annual increase in rent level has become the only variable controlled by housing policy; it is announced monthly in the news and on the official websites of government institutions. The defined increase in the rent level was not sufficient to regulate the private rented sector and ease the burden of housing expenditures for tenant households in Turkey. Because their owners' lease dwelling units for rental income, their priority is to gain extra profit. For a long duration of the tenancy, an owner can argue that the dwelling unit's rent level has remained below that of conjugate dwelling units due to the limit on the rent increase. When leasing is agreed upon for a tenancy period, the owner has a right to ask the tenant to leave the dwelling unit at the end of that period with a legal notice two months before the end of the contract period.

The amount of rent is indicated in the leasing agreement and the increase every year is determined by the annual change in price indexes. However, the owner decides on the difference in rent between current and future agreements (current and future tenants). This means that an owner can increase the rent for the future tenant as much as he/she likes after the current tenant leaves the dwelling unit. Therefore, the mobility of tenant households can affect the level of burden of housing and transport expenditures. Above all, the spatial distribution of rental units depends on the owner's decisions. When an owner decides to sell or not to rent a housing unit, there is no obligation to ensure the continuity of rental housing stock and vice versa. Hence, they are not provided in some locations. This situation can also be interpreted as the rental stock not being created according to any need or demand. It does not present a variety in size, internal organization, or physical structure. Although they exist, the diversity of rental housing units is very limited. All these restrict the availability of rental housing stock in

Turkey and lead to a mismatch between the size and location of housing units and their residents.

Extraordinary conditions are only defined for agricultural rents. A tenant is expected to pay less than the defined rent after it is agreed with an owner if unexpected weather occurs. However, this is not the case for residential usage. Therefore, many tenants and landlords (private and institutional owners of housing units) decided on housing rent payment if a unit is not lived in during the Covid-19 pandemic.

Lastly, according to Flat Ownership Law, tenants must take care of the leading property and preserve the property's architectural status, beauty, and strength. All costs except for monthly service dues are considered the responsibility of the owner of the housing unit. If an owner requests from the tenant of the housing unit to pay any additional cost for the dwelling unit and block of flats, the tenant has the right to cut that expenditure from their rent payments.

Observing the Burden of Housing and Transport Expenditures of Tenant Households

This chapter attempts to reveal items affecting the burden of housing and transport expenditures of tenant households in Turkey. Household Budget Survey-2018 cross-sectional data produced by the Turkish Statistical Institute are utilized. Additionally, the Household Budget Survey for the years 2008 and 2013 are referred to. First, an overall evaluation is made of the change in the modes of tenure and the demographic characteristics of these households are summarized for 2008, 2013, and 2018 (TURKSTAT, 2008, 2013, 2018). Then, to reveal the factors influencing the burden of housing and transport expenditures, 23 transport and 15 housing expenditure items are included. Housing expenditures are evaluated in three expense types: (i) housing rent, (ii) bills (water, electricity, heating), and (iii) basic maintenance, subscription, and service costs. Transport expenditures are categorized as public transport cost (passenger transport by train, underground, tram, bus, flights, sea vehicles, inland waterway, compound passenger transportation, cable railway, cable car, and chairlift transports) and private transport cost (first- and second-hand motorized and non-motorized vehicles, their tyres, spare parts and accessories, fuels, oils, maintenance and repair costs, rental and usage cost of garages, parking spaces and personal transportation vehicles and taxi costs, tolls and parking meters, and transport and shipping services).

Households are examined to observe the change in tenure structure, as shown in Table 14.1. Households living in housing units provided by the government/employers are not included due to the very low proportion in total. Owner-occupier, tenant, and others (living in their families' and relatives' housing unit with no rent or below market value) are focussed on. Significant changes in the share of the owner-occupier and other categories

Table 14.1 Descriptive characteristics of the modes of tenure in Turkey except for households living in a housing unit of the government/employers

Mean values for modes of tenure	Owner-occupier			Tenant			Others		
	2008	2013	2018	2008	2013	2018	2008	2013	2018
Share in that year (%)	66	61.5	60.7	22.6	21.9	24.4	10	14.2	14.3
Size of hh (persons)	3.9	3.8	3.5	3.7	3.4	3.3	3.7	3.5	3.3
Age of hh head (years)	na	52.8	54.8	na	40.5	42.5	na	44.8	47.1
Occupancy duration (years)	15.9	16.9	19	4.5	3.9	4.7	11.7	15.2	15.7
Annual income (TL)	18,589	30,150	53,253	17,482	29,276	52,288	10,312	21,961	46,525
Number of bedrooms	3.5	3.6	3.6	3.4	3.4	3.4	3.3	3.3	3.4

Source: Prepared by the authors based on the Household Budget Survey-2008–2013–2018 cross-sectional raw data.

have been observed; in contrast, tenants have remained constant. In line with Whitehead et al.'s (2012) findings for the many European countries, households living in the private rented sector are usually smaller in size and younger and have lower incomes than owner-occupiers in Turkey. The others category has remained between the owner-occupier and tenant households in terms of these characteristics. Regarding occupancy duration, owner-occupier households and others have higher mean values than tenant households. Lastly, there is no significant difference between the number of bedrooms in a housing unit in which owner-occupier, tenant, and others households live.

The housing and transport expenditures burden of owner-occupier households is not observed because of the lack of input for housing debt and credit payments of these households. Similarly, although others households indicate they pay some rent to their relatives or family, the size of this payment is not recorded. This chapter examines two arguments explaining the housing and transport expenditures burden of tenant households: the mobility of tenant households and the adequacy of the current unit for the residing tenant household. In that respect, the mobility of households is defined dichotomously through the median occupancy years of tenant households in their current unit: above (four years or more) and below (three years or less) the median. The adequacy of the housing unit is defined by having a sufficient number of bedrooms for every household member; yet the need of couples is accepted as one bedroom. As a result, tenant households are grouped in three categories: having adequate bedrooms, having inadequate bedrooms (less than they need), and having excessive bedrooms (more than they need).

In evaluating tenant households' burden status category (see Table 14.2), housing rent is observed to take up the highest share of housing and transport expenditures. Although the shares of bills; basic maintenance, subscription and service costs; and public transport in the household budget do not distinguish households experiencing burden from those with no burden, the shares of housing rent and private transport expenditures vary. Like the burden status, the share of housing rent in the household budget changes with the occupancy duration of tenant households. Households who have lived for three years or less in the current rented housing unit make up a higher share than those residing there for four years or more. The percentage of housing rent in the household budget varies with the adequacy of bedrooms for the residing tenant household; households with an inadequate number of bedrooms pay a lower share of housing rent but a higher share for bills than the other two categories. There is no prime difference in the share of housing rent in the household budget for households with an adequate or excessive number of bedrooms.

Rather than a descriptive interpretation, a t-test is applied to categories (burden status and occupancy duration) and one-way ANOVA to the adequacy of bedrooms' category to determine whether there is a statistically significant difference between the sub-categories of each category in terms

Table 14.2 The share of housing and transport expenditure items of tenant households in burden status, occupancy duration, and adequacy of bedrooms

Category	Sub-category	Housing rent	Bills	Basic maintenance, subscription, and service cost	Private transport	Public transport	Total housing and transport expenditures
Burden	No	0.1319	0.0434	0.0005	0.0201	0.0042	0.2000
	Yes	0.2914	0.0950	0.0010	0.1034	0.0092	0.5000
Occupancy duration	3 years or less	0.2049	0.0609	0.0007	0.0504	0.0067	0.3237
	4 years or more	0.1750	0.0642	0.0006	0.0514	0.0052	0.2964
	Inadequate	0.1753	0.0700	0.0004	0.0508	0.0059	0.3025
Adequacy of bedrooms	Adequate	0.2016	0.0623	0.0008	0.0548	0.0056	0.3252
	Excessive	0.1981	0.0547	0.0009	0.0478	0.0065	0.3080

Source: Prepared by the authors based on the Household Budget Survey-2018 cross-sectional raw data.

of the share of housing rent; bills; basic maintenance, subscription, and ser-vice costs; private transport; and public transport in the household's budget. Since $p < .001$, i.e., less than the significance level $\alpha = 0.05$ for Levene's test for equality of variances, the null hypothesis (there is no significant differ-ence between two sub-categories of burden status and occupancy duration) is rejected. Significant differences are shown in bold in Table 14.2. The results indicate that the mean of the share of housing rent ($t_{986.656} = -20.819$, $p < .001$); bills ($t_{1114.571} = -20.893$, $p < .001$); basic maintenance, subscription, and ser-vice costs; ($t_{1142.981} = -4.854$, $p < .001$); private transport ($t_{949.528} = -9.981$, $p < .001$); and public transport ($t_{1040.121} = -6.206$, $p < .001$) for households having a bur-den and those with no burden is significantly different. Households having a burden of housing and transport expenditures spend more of their income on these items.

For the occupancy duration category, the means of the shares of housing rent ($t_{2117.547} = 4.725$, $p < .001$) and public transport ($t_{2392.872} = 2.486$, $p < .001$) in the household budget are statistically significant. Households who have lived for three years or less in their current rented units spend a higher pro-portion of their incomes on housing rent and public transport.

It is concluded that the mean values of the share of housing rent ($F_{2,2522} = 6.478$, $p < 0.001$), bills ($F_{2,2522} = 17.300$, $p < 0.001$), and basic maintenance, subscription, and service costs ($F_{2,2522} = 12.071$, $p < 0.001$) differ significantly between households having different adequacy levels of bedrooms in the housing unit. Compared to households with inadequate bedrooms, those with adequate or excessive numbers of bedrooms spend a high share of their incomes on housing rent and basic maintenance, subscription, and service costs. However, the opposite is seen for the percentage spent on bills.

In determining the role of household income in the share of housing and transport expenditures, households are assessed in terms of their equalized household income. Before this assessment, it would be helpful to summa-rize the distribution of income groups among tenant households. The lowest income group in tenant households makes up 18.4% of tenant households and the highest income group is 23.7%. Although the housing rent is higher for higher-income households, the share of housing rent in household income is less than in lower-income groups. Moreover, 18.5% of the highest income tenant households are experiencing the burden of housing and transport expenditures; it reaches 63.7% of lowest income group tenant households. In that respect, a higher income level is associated with a lower share of housing and transport expenditures.

Lastly, this chapter's two arguments (mobility and adequacy) focus on the division of households having a burden of housing and transport expendi-tures and those not having one. This is represented in Table 14.3, where the shares of all tenant households are shown without parentheses, with those of the lowest income tenant households in parentheses. While 51.5% of ten-ants having no burden of housing and transport expenditures have been residing in their current unit for three years or less, 56.5% of tenants having

Table 14.3 Cross-tabulations of burden status with occupancy duration and adequacy of bedrooms

Category	Sub-category	Burden	
		No	*Yes*
Occupancy duration	3 years or less	51.5 (45.8)	56.5 (55.3)
	4 years or more	48.5 (54.2)	43.5 (44.7)
Adequacy of bedrooms	Inadequate	36.8 (74.1)	34.6 (60.1)
	Adequate	29.1 (19.9)	28.6 (26.1)
	Excessive	34.1 (6.0)	36.8 (13.8)

Source: Prepared by the authors based on the Household Budget Survey-2018 cross-sectional raw data.

a burden have been living there for three years or less. However, it decreases to 45.8% of the lowest-income tenant households having no burden and residing there for three years or less. As a result, the findings support this chapter's arguments: if a tenant household has lived for a shorter duration in the current housing unit, they are more frequently facing a burden of housing and transport expenditures.

Living in a housing unit whose number of bedrooms is adequate for household composition means a higher chance of not facing the burden of housing and transport expenditures. For example, 28.6% of households having a burden live in a housing unit with an adequate number of bedrooms. Yet, it reaches 34.6% for those having an inadequate number of bedrooms and 36.8% for those having an excessive number of bedrooms. For lowest-income households, whether they have a burden or not, the share of households with an inadequate number of bedrooms is much higher (74.1%) than that of households with an adequate (19.9%) and an excessive number of bedrooms (34.1%). In the examination of households having no burden but an inadequate number of bedrooms, these households are observed to have higher numbers of members (4.79 people in a household), with 2.97 people for the remaining tenant households and 3.4 people for all tenant households. Although the level of the total income of these households (no burden but an inadequate number of bedrooms) is higher than that of the remaining households and all tenant households, the value of equalized household income is lower due to the high household size.

Discussion and Conclusion

In this chapter, housing and transport expenditure items of Turkish tenant households are examined for 2018. As argued, adequacy of the number of bedrooms and mobility of tenant households are observed to be significant factors to determine the share of housing rent in the household budget. Longer duration of occupancy (four years or more) and an inadequate or excessive number of bedrooms are associated with a lower share of rent and, accordingly, housing and transport expenditures. These findings are valid for all income groups; however, the lowest-income households experience more of these expenditures than the other income groups. Being a lowest-income household also means an increase in the possibility of having an inadequate number of bedrooms and a decreased possibility of having an adequate and excessive number of bedrooms. Similarly, private transport expenditures determine the experience of the burden of housing and transport expenditures of tenant households. The lack of public transport facilities is most to blame, although the spatial expansion trend has been adopted in the urban development pattern of Turkey in recent years.

As a result of Turkey's current housing dynamics, based on producing new housing units at the urban peripheries and rebuilding existing housing stock through urban transformation projects, the need to regulate the private rented sector is evident to enable the maintainability of rentable housing stock and provide various price levels and diversity in size and location. Because more urban transformation projects are applied to areas with comparatively low rent housing units, the amount of affordable housing units will be too small to meet the population's needs. Households who do not want to be homeowners and/or are not capable of being so will possibly face higher housing expenditures (rent, subscription, etc.).

Following the discussion of the burden of transport and housing expenditures for tenant households, there are some points to mention about the deficiency and improperness of variables in the Household Budget Survey dataset. First of all, tenant households pay deposits for a rented housing unit in cash and instalments in Turkey. Mostly, it corresponds to the value of one month's rent. It aims to protect the owner after the tenancy has ended. If the owner finds any unexpected damage to the property, this deposit can be used for repairs. Otherwise, the owner has to pay the deposit back to the tenant. In addition to the deposit, if an estate agent has acted as mediator between tenant and owner, the tenant has to pay one month's rent to the agent as a commission fee. However, the Household Budget Survey does not include these expenditures as variables. Both increase the amount of the initial payment and lead to a burden of housing expenditures for tenant households.

Another drawback of the Household Budget Survey derives from the data gathering method. Households are asked to report their expenditures only for a given month that is observed by TURKSTAT. This causes some expenditures (such as transport and shipping services) to be neglected when a tenant does not move in the observed month. However, most of the housing units in Turkey are leased without furniture, which means each household has to move their furniture from a previous housing unit or storage to a newly rented unit.

References

Aksoy E (2017) Housing affordability of different income groups in Turkey: Regional comparison. Unpublished Master Thesis. Middle East Technical University, Ankara.

Aksoy Khurami E (2021) *Failing promises of homeownership in Turkey*. Unpublished PhD Thesis. Middle East Technical University, Ankara.

Alkan L, Uğurlar A (2015) Türkiye'de Konut Sorunu ve Konut Politikaları. Ankara: 72 Tasarım Dijital Basımevi.

Aşıcı M, Yılmaz Ö and Hepşen A (2011) Housing affordability index calculation integrating income inequality in Turkey. *Empirical Economics Letters* 10(4): 359–367.

Balamir M (1999) Formation of private rental stock in Turkey. *Netherlands Journal of Housing and the Built Environment* 14: 385–402.

Beer A, Kearins B and Pieters H (2007) Housing affordability and planning in Australia: The challenge of policy under neo-liberalism. *Housing Studies* 22(1): 11–24. DOI: 10.1080/02673030601024572.

Belsky ES, Goodman J and Drew R (2005) Measuring the nation's rental housing affordability problems. The Joint Center for Housing Studies, Harvard University. Available at: https://www.jchs.harvard.edu/sites/default/files/media/imp/rd051_measuring_rental_affordability05.pdf. Accessed 10 March 2021.

Bogdon AS, Can A (1997) Indicators of local housing affordability: Comparative and spatial approaches. *Real Estate Economics* 25(1): 43–80.

Bourassa SC (1996) Measuring the affordability of homeownership. *Urban Studies* 33(10):1867–1877.

Bramley G, Munro M and Pawson H (2004) *Key Issues in Housing: Policies and Market in 21st Century in Britain*. London: Palgrave Macmillan.

Byrne M (2019) Generation rent and the financialization of housing: A comparative exploration of the growth of the private rental sector in Ireland, the UK and Spain. *Housing Studies* 35(4): 743–765. DOI: 10. 1080/02673037.2019.1632813.

Collinson R (2011) Rental housing affordability dynamics, 1990–2009. *Cityscape: A Journal of Policy Development and Research* 13(2): 71–103.

Coşkun Y (2020) Measuring homeownership affordability in emergent market context: An exploratory analysis for Turkey. *International Journal of Housing Markets and Analysis 14(3): 446-480*. DOI: 10.1108/IJHMA-04-2020-0033.

Dewilde C (2018) Explaining the declined affordability of housing for low-income private renters across Western Europe. *Urban Studies* 55(12): 2618–2639.

Dodson J, Li T and Taylor E et al. (2020) Community burden and housing affordability for low-income renters. AHURI Final Report No. 335. Melbourne: Australian Housing and Urban Research Institute Limited. Available at: https://www.ahuri.edu.au/research/final-reports/335. Accessed 10 February 2021.

Edgar B, Doherty J and Meert H (2002) *Access to housing: Homelessness and vulnerability in Europe*. Bristol: Policy Press.

Emekci Ş (2018) A life cycle costing based decision support tool for cost-optimal energy efficient design and/or refurbishments. Unpublished PhD Thesis. Middle East Technical University, Ankara.

Fields D (2018) Constructing a new asset class: Property-led financial accumulation after the crisis. *Economic Geography* 94(2): 118–140.

Galster G, Lee KO (2021) Housing affordability: A framing, synthesis of research and policy, and future directions. *International Journal of Urban Sciences* 25(1): 7–58. DOI: 10.1080/12265934.2020.1713864.

Haffner MEA, Hulse K (2021) A fresh look at contemporary perspectives on urban housing affordability. *International Journal of Urban Sciences* 25(1): 59–79. DOI: 10.1080/12265934.2019.1687320.

Hulchanski JD (1995) The concept of housing affordability: Six contemporary uses of the housing expenditure-to-income. *Housing Studies* 10(4): 471–492.

Jewkes MD, Delgadillo LM (2010) Weaknesses of housing affordability indices used by practitioners. *Journal of Financial Counseling and Planning* 21: 43–52.

Joyce R, Mitchell M and Norris Keiller A (2017) *The Cost of Housing for Low-income Renters*. London: Institute for Fiscal Studies.

Krapf S, Wagner M (2020) Housing affordability, housing tenure status and household density: Are housing characteristics associated with union dissolution? *European Journal of Population* 36: 735–764.

Leishman C, Rowley S (2012) Affordable Housing. In: Clapham DF, Clark WAV, Gibb K (eds) *The Sage Handbook of Housing Studies*. London: Sage Publications, 379–396.

Leone R, Carroll BW (2010) Decentralisation and devolution in Canadian social housing policy. *Environment and Planning C: Government and Policy* 28: 389–404.

Malpass P, Murie M (1999) *Housing Policy and Practice*. Basingstoke: Macmillan.

Nassi CD, da Costa FCDC (2012) Use of the analytic hierarchy process to evaluate transit fare system. *Research in Transportation Economics* 36: 50–62. DOI: 10.1016/j.retrec.2012.03.009.

Özdemir Sarı ÖB, Aksoy E (2016) Excess production, rising prices, and declining affordability: Turkish housing experience. In: Wroot I (ed.) *AMPS Conference Publication Series 8. Government and Housing in a Time of Crisis: Policy, Planning, Design and Delivery*. Liverpool: Liverpool John Moores University, 162–170.

Özdemir Sarı ÖB, Aksoy Khurami E (2018) Housing affordability trends and challenges in the Turkish case. *Journal of Housing and the Built Environment*. DOI: 10.1007/s10901-018-9617-2.

Padley M, Marshall L and Valadez-Martinez L (2019) Defining and measuring housing affordability using the Minimum Income Standard. *Housing Studies* 34(8): 1307–1329. DOI: 10.1080/02673037.2018.1538447.

Preece J, Hickman P and Pattison B (2020) The affordability of "affordable" housing in England: Conditionality and exclusion in a context of welfare reform. *Housing Studies* 35(7): 1214–1238. DOI: 10.1080/02673037.2019.1653448.

Quigley JM, Raphael S (2004) Is housing unaffordable? Why isn't it more affordable? *Journal of Economic Perspectives* 18(1): 191–214.

Revington N, Townsend C (2016) Market rental housing affordability and rapid transit catchments: Application of a new measure in Canada. *Housing Policy Debate* 26(4–5): 864–886. DOI: 10.1080/10511482.2015.1096805.

Sarıoğlu-Erdoğdu GP (2014) Housing development and policy change: What has changed in Turkey in the last decade in the owner-occupied and rented sectors? *Journal of Housing and the Built Environment* 29: 155–175.

Sarıoğlu-Erdoğdu, GP (2015) Well-being of renters in Ankara: An empirical analysis. *Habitat International* 48: 30–37.

Sendi R (2014) Housing accessibility versus housing affordability: Searching for an alternative approach to housing provision. *Sociologija I Prostor* 200(3): 239–260.

Singer ME (2020) How affordable are accessible locations? Housing and transportation costs and affordability in U.S. Metropolitan areas with intra-urban rail service. PhD dissertation. The University of Michigan, Michigan.

Stone ME (1993) *Shelter Poverty: New Ideas on Housing Affordability*. Philadelphia: Temple University Press.

Stone ME (2006) What is housing affordability? The case for the residual income approach. *Housing Policy Debate* 17(1): 151–184.

Thalmann P (1999) Identifying households which need housing assistance. *Urban Studies* 36(11): 1933–1947.

TURKSTAT (2004) *Household Budget Survey-Micro Data Set*. Ankara: Turkish Statistical Institute.

TURKSTAT (2008) *Household Budget Survey-Micro Data Set*. Ankara: Turkish Statistical Institute.

TURKSTAT (2013) *Household Budget Survey-Micro Data Set*. Ankara: Turkish Statistical Institute.

TURKSTAT (2018) *Household Budget Survey-Micro Data Set*. Ankara: Turkish Statistical Institute.

Uğurlar A, Özelçi Eceral T (2017) Özel Kiralık Konut Sektörü ve Politikaları: Dünyadan Farklı Yaklaşım ve Düzenleme Örnekleri. *Trakya Üniversitesi Mühendislik Bilimleri Dergisi* 18(1): 53–71.

Walks A (2018) Driving the poor into debt? Automobile loans, transport disadvantage, and automobile dependence. *Transport Policy* 65: 137–149. DOI: 10.1016/j.tranpol.2017.01.001.

Whitehead C, Monk S and Markkanen S et al. (2012) *The Private Rented Sector in the New Century: A Comparative Approach*. Copenhagen: Boligøkonomisk Videncenter.

Yates J (2006) Housing affordability: What it is and why does it matter? Paper presented to the Australasian Housing Researchers Conference, Adelaide.

Yates J, Gabriel M (2006) Housing affordability in Australia. *AHURI Research and Policy Bulletin* 68. Available at: https://apo.org.au/sites/default/files/resource-files/2006-04/apo-nid4593.pdf. Accessed 5 February 2021.

15 Urban Transformation Policies of the 21st Century
State-Led Gentrification

Nil Uzun

Introduction

Gentrification has been an important area of research and theoretical debate since it was first defined by Ruth Glass in 1964. When the theoretical explanations and the results of field studies are evaluated together, three basic conditions show the presence of gentrification in an area. First of all, a previously depreciated housing stock, mostly in a central part of the city, is renewed or redeveloped and gains value. Second, the high-income groups displace the low-income groups. The last condition is that the ownership pattern changes in the area and a transformation from tenancy to owner-occupation occurs. As a result, the physical structure of the old residential areas changes and the quality of the houses and the service offered increases (Uzun, 2001). Gentrification is among the local consequences of a multidimensional and interrelated network that brings about economic, socio-cultural, and spatial transformations in urban spaces (Van Weesep, 1994). This process, which initially emerged via individual initiatives, is shaped by state intervention at both national and local levels in the later stages, and in the 21st century, the nature of the process has changed with the effect of market dynamics.

While studies on gentrification in the United States and Europe started in the 1970s, research on the subject in Turkey came to the fore at the end of the 1980s, when the first examples of gentrification were observed in Istanbul. At the beginning of the 21st-century neoliberal economic policies matured and neoliberal restructuring that has been going on since the 1980s spread to all social and spatial relations with new political and economic actors, leading to new forms of socio-spatial segregation. The number and impact of urban transformation projects have also increased and state-led gentrification has become a part of urban transformation policies. In this section, the state-led gentrification in Turkey is discussed with an example from Istanbul.

State-Led Gentrification

There have been many theoretical and empirical studies on gentrification since its first definition in 1964 (Glass,1964). As the number of the relevant

DOI: 10.4324/9781003173670-19

examples increased, gentrification turned into a global phenomenon and became a global urban strategy (Smith, 2002). Therefore, it is no longer experienced only in world cities undergoing important reinvestment in their central districts but also in other cities around the world. It has also been a widespread part of urban planning policy in many cities. Based on the extensive research on gentrification, Slater's (2009: 294) definition appears to be interpretative and up to date. He defines gentrification as "... the transformation of a working-class or vacant area of a city into middle-class residential and/or commercial use". He adds the vacant areas to his definition "because of the many instances of exclusive 'new build' gentrification, which often occur on formerly working-class industrial spaces".

After nearly 40 years of research and theoretical work, Hackworth and Smith (2001) described three distinct waves of global gentrification at the beginning of the 21st century. These waves were identified by the observation of gentrification in New York City. In the first wave, which took place in the five-year period from 1968 onward, gentrification was isolated in small neighbourhoods. This was followed by a transition period in which the gentrifiers acquired property. Developers and investors, making use of the downturn in property values, bought up large portions of land in depreciated neighbourhoods. This also set the stage for the second wave occurring in the 1980s. In the second wave, gentrification became a common process seen also in smaller, non-global cities. Moreover, there were intense political struggles over the displacement of the poorest residents. Another transition period in which gentrification declined was witnessed between the end of the 1980s and the first years of the 1990s. In the last wave, starting in the mid-1990s, while gentrification continued in many inner-city neighbourhoods, it was also observed in neighbourhoods far from the city centre for the first time. Furthermore, large-scale developers transformed neighbourhoods as a whole, often with government support. Therefore, gentrification became more linked to large-scale capital than ever before, and instead of individual pioneers, the leading initiators of gentrification became corporate developers. There are several features that distinguish this state-led gentrification wave from previous waves. This time gentrification is triggered by large companies and state actors at an unprecedented rate. Moreover, anti-gentrification movements are becoming increasingly ineffective and the process tends to spill over into more decentralized areas that have not felt the pressure of gentrification before (Hackworth and Smith, 2001; Hackworth, 2002).

In the early 2000s, at the national or global level, capital arrived in the depreciated inner-city areas in advantageous locations, turning them into investment opportunities with the support of the state. Following these developments, Lees, Slater, and Wyly (2008) defined a fourth wave of gentrification, stating that with the strong integration of local gentrification with national and global capital markets, more state policies began to encourage gentrification. This fourth wave is considered specific to the US because it "is not readily identifiable outside of the United States" (Lees et al. 2008: 184).

Aalbers (2019: 5) argues that "fourth-wave gentrification is a continuation or even intensification of third-wave gentrification". Following Lees, Slater, and Wyly (2008), Aalbers (2019: 1) stated that "... during the global financial crisis we have entered fifth-wave gentrification. Fifth-wave gentrification is the urban materialisation of financialised or finance-led capitalism. The state continues to play a leading role during the fifth wave, but is now supplemented – rather than displaced – by finance". Although the fourth and fifth waves have been experienced and defined recently, together with the third wave, they are expressions of the economic conditions and processes making reinvestment in disinvested inner-urban areas very attractive for investors. Therefore, this time gentrification developed mostly in relation to big capital. Large-scale developers transformed neighbourhoods as a whole, frequently benefitting from government support. To summarize, after the 1990s, the role of the state in gentrification differed in the last three waves and state-led gentrification became the dominant policy. As Slater (2006: 749–750) summarizes, "The current era of neoliberal urban policy, together with a drive towards homeownership, privatization and the break-up of 'concentrated poverty' (Crump, 2002), has seen the global, state-led process of gentrification via the promotion of social or tenure 'mixing' (or 'social diversity' or 'social balance') in formerly disinvested neighbourhoods populated by working-class and/or low-income tenants (Hackworth and Smith, 2001; Smith, 2002; Slater, 2004, 2005; plus many articles in Atkinson and Bridge, 2005)". During this process, urban transformation is possible through changes in urban policy such as public–private partnerships, developer subsidies, and urban place marketing. Initially gentrification was a gradual process, but in the last three decades, it has evolved into a more coordinated state-led process transforming neighbourhoods entirely and rapidly (Gordon et al., 2017). In the 21st century, the neoliberal state no longer takes the stage as a regulator but as an active actor in the market. In this period, there is a significant relationship between gentrification and urban transformation projects and neoliberal policies. Urban transformation represents the next wave of gentrification, planned and financed on an unprecedented scale. Although gentrification is often concealed beneath the cloak of urban transformation, there are explicit or covert gentrification intentions behind many urban transformation projects. In addition, state-led gentrification is marketed under the name neighbourhood transformation (Smith, 2002; Lees and Ley, 2008). Researchers focussing on gentrification emphasize that it is not correct to make a sharp distinction between urban transformation processes and gentrification. The general tendency in this regard is to interpret these two processes as a whole (İslam, 2009).

State-Led Gentrification in Turkey

In line with the aim of integration with the global market in the 1980s, many metropolitan cities in Turkey entered a period of significant transformation as a result of both local and international dynamics. Changes such as the partial independence of municipalities from the central administration, their gaining entrepreneurial potential in connection with the changes in the world system, the emergence of large-scale projects, the increase in housing investments, the change in consumption patterns, the adaptation of the production system to the worldwide flexible capital flow system at a certain level, and the expansion of the service sector led to the transformation of the urban space. In addition, migration from rural areas to metropolitan cities continued and the migrating population settled in squatter housing areas as well as in vacant or outdated houses in the central neighbourhoods of cities. Gentrification began in these inner-city areas with historical, architectural, and aesthetic value that had become increasingly dilapidated between the 1960s and 1980s. At the end of the 1980s, examples of gentrification were observed for the first time, especially in Istanbul (Ergün, 2006; Sönmez and Geniş 2013; Uzun, 2015). Gentrification, initially appearing only as small-scale improvement and transformation, has been turned into a comprehensive restructuring of the squatter housing areas by the partnership between the state and private sector since the beginning of the 21st century and the number of examples has increased rapidly.

Gentrification in Turkish cities like Istanbul, Ankara, and İzmir can be interpreted in terms of the waves of gentrification and Slater's (2009) definition mentioned in the previous section. The Turkish cases have similarities to other examples from around the world. The waves of gentrification have been observed in different neighbourhoods in similar patterns but in time periods different from those defined by Hackworth and Smith (2001). Like other cities around the world, large-scale private investors were involved in the third wave of gentrification. Moreover, state-led urban transformation projects have led to the forced movement of the urban poor out of the central parts of the city, resulting in profits being made from the land and property market and state-led gentrification. Along with the problems of transforming the city centre, planners and policymakers have had to deal with the impact of changes in the residential areas of cities. Increasingly, urban transformation projects have been their main tools for controlling the transformation of residential areas, especially in squatter housing areas. These projects have attempted to turn squatter housing neighbourhoods into residential areas with high environmental quality and better living conditions. However, most of these, especially those in the last two decades, resulted in the displacement of squatters and provision of luxurious housing for high-income groups, leading to state-led gentrification. Therefore, the Turkish urban transformation experience over the last two decades can be interpreted based on the explanations by Lees and Ley (2008) and Smith (2002), stating that urban transformation represents state-led gentrification and it is marketed under the name neighbourhood transformation. These urban transformation projects have been declared solutions to all urban problems such as earthquakes, crime, segregation, and poor living conditions. Eventually, they serve the appropriation of existing land for the use of high-income groups, giving rise to state-led gentrification. In a similar vein, the old squatter housing areas, now enclosed in the inner sections of the city, have rapidly been transformed into upper middle class and middle class residential areas by large-scale urban transformation projects (Özdemir, 2010; Çavdar and Tan, 2013; Türkün, 2014; Uzun, 2015, 2020).

Urban Transformation Projects and State-Led Gentrification

In Turkey, policies supporting the construction sector after the 2001 economic crisis led to the acceleration of urban transformation projects and related legal regulations. Urban transformation projects have occupied an important place on the agenda of the government since 2000 and were implemented in several parts of the cities. As historic areas in the central parts of cities gained great importance over the last 40 years, preserving and restoring historical buildings and often changing their functions and reusing them form the basis of the urban transformation projects in these areas. In addition, squatter houses and neighbourhoods in the city centre have been where most urban transformation has taken place. Another area where transformation has taken place is the parts of the city with old industrial buildings. These areas have been renewed and offered for commercial and cultural uses as well as residential uses. Especially after the 1999 Marmara earthquake in Turkey, urban transformation projects in the areas with earthquake risk have also come to the fore. As a result, urban transformation projects have turned into an important tool for local governments. Urban transformation projects are of great importance for improving the areas subject to transformation and facilitating the production of healthier living environments. Conversely, rather than aiming for long-term improvement of the economic, social, physical, and environmental structure of the transformation areas, projects have started to become a source of income for local governments and different interest groups and mostly resulted in state-led gentrification.

At the beginning of the 21st-century urban transformation projects have been promoted by several pieces of legislation. These have provided the Housing Development Administration (TOKİ), Ministry of Environment and Urbanization, and the local governments with new powers to implement urban transformation projects, which led to state-led gentrification. The first of the legal regulations for urban transformation during this period is the North Ankara Entrance Urban Transformation Project Law (Law no. 5104), which was enacted in 2004. The purpose of this law is to increase the quality of urban life by improving the physical conditions and environmental image, beautifying, and providing a healthier settlement order within the framework of the urban transformation project in the areas covering the northern entrance to Ankara and its surroundings. With this law, the boundary of the urban transformation project area is determined and the planning and implementation authority of the district municipalities in the project area is transferred to the Ankara Metropolitan Municipality. The project area was a squatter housing area containing 10,500 such houses and they were replaced by high-rise blocks of flats. The project was carried out in partnership with the Ankara Metropolitan Municipality and TOKİ and only a physical transformation was envisaged in the project (Korkmaz, 2015). After the demolition of the squatter houses in the project area, a transformation took place by which the population density increased with the high-rise construction. The beneficiaries and tenants in the area, on the other hand, have been placed in new blocks of flats built by TOKİ in Karacaören, 5 km from the project area. The squatters who were able to provide proof of ownership in the project area on a date earlier than 1 January 2000 are determined as beneficiaries and have the opportunity to own one of these dwellings. Otherwise, squatters are given a chance to buy a dwelling unit via a ten-year payment plan (Korkmaz, 2015). In addition, five types of beneficiaries have been defined within the scope of the project. Agreements between beneficiaries and TOKİ are made and, depending on the type and size of the property beneficiaries own in the project area, reimbursements for the new dwellings are determined (Yüksel, 2007). With this urban transformation project, which is an example of state-led gentrification, a squatter housing neighbourhood was turned into a mass housing site, leading to social segregation.

In the Municipality Law (Law no. 5393) enacted in 2005 regarding urban transformation projects, it is stated that municipalities may implement urban transformation and development projects in order to rebuild and restore the old parts of the city in accordance with the development of the city; create housing areas, industrial and commercial areas, technology parks, and social facilities; take measures against earthquake risk; or protect the historical and cultural texture of the city. Although the areas where urban transformation projects can be implemented have been defined in a relatively comprehensive way, a numerical restriction has been imposed only on the size of the area in order for an area to be declared an urban transformation and development project area. There is no comprehensive foresight about how the transformation process will take place. In the Metropolitan Municipality Law (Law no. 5216), the same authority is given to the Metropolitan Municipalities. In line with these two laws, municipalities declare especially the areas where squatter houses are located urban transformation areas and carry out transformation. However, sometimes municipalities transfer their authority in this area to TOKİ after declaring the urban transformation area, and, as a result, TOKİ carries out the transformation. The squatter housing neighbourhoods that are located in the favourable locations of cities are declared urban transformation project areas with reference to these laws. In most cases, these transformation projects result in displacement of squatters and the transformation of the area into a high-income neighbourhood and therefore give rise to state-led gentrification.

Another law enacted in 2005 (Law no. 5366) is entitled "Renewal, Protection and Use of Worn-out Historical and Cultural Immovable Properties". The purpose of this law is to reconstruct and restore dilapidated areas that are about to lose their characteristics and that have been registered and declared protected areas by the cultural and natural heritage conservation boards and the protection areas in these regions, in accordance with the development of the region. It is also aimed to create residential, commercial, cultural, tourism, and social amenity areas in these regions; to put in place measures to counter the risk of natural disasters; and to renew and preserve historical and cultural immovable assets and keep them alive by using them. This purpose coincides with the objectives of urban transformation projects covering historical urban textures. With the enactment of this law, districts located in the historical city centres and whose rent value has increased have also been declared urban transformation areas. This law draws a different legal framework for places declared urban transformation areas and property owners are forced to choose one of the options offered to them within the context of the urban transformation project. One of the most important decisions contained in this law to facilitate transformation is the urgent expropriation authority provided to the local governments that can be used in the event that a compromise with the property owners cannot be reached. Another is the regulations that allow municipalities to carry out projects on a neighbourhood scale in protected areas. In this way, areas where urban transformation was not possible spontaneously until that time transform rapidly with the new authority given to municipalities and state-led gentrification takes place. The first noteworthy implementation of Law no. 5366 was realized in Istanbul with the Sulukule urban transformation project that will be explained in the next section (Türkün and Sarıoğlu, 2014).

The latest law regarding urban transformation (Law no. 6306) was enacted in 2012. This law, entitled "Transformation of Areas under Disaster Risk", determines the procedures and principles for improvement, clearance, and transformation in order to create healthy and safe living environments in areas at risk of disasters and in areas with risky structures. The Ministry of Environment and Urbanization is designated as the responsible ministry. With this law and an amendment made to the article of the Municipality Law on urban transformation in 2010, a wide implementation area has been provided to TOKİ and important authority has been given to transform the squatter housing areas completely. This authorization is valid in squatter housing neighbourhoods, for public land owned by TOKİ, and in areas declared mass housing areas. As one of the common urban transformation practices of this period, the low-income groups living in central parts of cities that have begun to deteriorate are displaced to mass housing areas constructed by TOKİ in locations far from the city centres. The authorization granted to TOKİ by various legal regulations to develop projects directly or through subsidiaries within the country or abroad and to directly lend to projects for the transformation of squatter housing areas and the transformation of historical areas has led to TOKİ's playing an important role in the state-led gentrification process (İslam, 2009).

Sulukule Urban Transformation Project

Sulukule is an ancient settlement located right on the shore of the historical land walls that form the boundaries of the historical peninsula in Istanbul. Located within the borders of two neighbourhoods, the project area consists of mostly detached row houses with 1–2 floors and blocks of flats with 4–5 floors in some places. Sulukule was associated with the Roma who lived there for decades and the entertainment venues they operated in the past, which occupied a certain place in the entertainment life of the city. In the 1990s, the region entered a period of decay with the closure of the entertainment venues, which constitute the main economic activity in the region and provide a source of income for many families through direct and indirect jobs. The region was a cheap housing stock for the poor with property and rent values that were well below market averages until recently (İslam, 2009). It also represents a typical inner-city poverty zone (Ünsal, 2013).

The transformation process in Sulukule was initiated immediately after the enactment of Law no. 5366. First, a protocol was signed between the Fatih Municipality, TOKİ, and the Istanbul Metropolitan Municipality in 2005. The Istanbul Metropolitan Municipality functions as the umbrella institution that supervises urban transformation, whereas the Fatih Municipality acts as the local coordinator. This district municipality is responsible for identifying the tenure structure of the neighbourhood, pursuing negotiations with property owners, and undertaking expropriation when needed. TOKİ acts as the project developer. It is responsible for preparing the plans for the area as well as handling the construction of new units. The project foresees the demolition of existing buildings and the construction of new buildings by TOKİ in their place. In the project prepared, there is a hotel, school, and cultural facility beside the housing units. The project directly affected approximately 3500 families (İslam, 2009; Ünsal, 2013).

Two options were offered to those who owned property in the area declared Sulukule Urban Transformation Area. The first option is to be involved in the project prepared within the framework of the conditions determined by the municipality. According to the protocol of the project, the property owners are given the right to own the houses to be built, provided that they pay the difference between the construction costs and the current values of the houses in monthly instalments over up to 15 years. The second option is to give the property to the municipality in return for money or barter (housing in another region). In the event these two options are not accepted, the residents are informed that expropriation will be resorted to. Therefore, property owners were not offered the chance to opt out of the project. This has caused some of the property owners to sever their connections with the area by selling their properties to third parties who bid higher for their properties than the municipality. On the other hand, 62% of the property owners were unable to consider being registered for the new houses planned to be built in Sulukule as their financial situations did not allow them to pay the instalments. The tenants living in the region have been granted the right to own a house in the mass houses built by TOKİ in Taşoluk (a neighbourhood roughly 40 km from the project area) with a 2000 TL prepayment and monthly instalments of 300–400 TL spread over 15 years. However, 37% of the tenants did not register for the houses in Taşoluk due to the repayment conditions, which they could not afford. With the expropriation decision made in December 2006, the demolition of the neighbourhood was authorized. Approximately 500 families were settled in Taşoluk. During the process, the tenants were also permitted to transfer their rights. After the right of transfer, a significant portion of the tenants, the majority of whom do not have a regular job or income and have difficulty in paying the extra costs incurred in new settlements (natural gas connection costs, monthly natural gas bills, building dues, transportation, etc.), sold these rights to third parties. Even before the transfer right was granted, flats began to change hands informally in exchange for 3000–5000 TL. Following the granting of the transfer right and the official realization of the transfer of rights, the values of these flats reached between 20,000 and 40,000 TL, depending on variables such as location, size, and date of sale. Only a small portion of the people living within the boundaries of the project area have a regular income. Therefore, for a significant number of the homeowners, staying in the area after the project means finding new sources of income in order to be able to pay the monthly instalments of the houses to be built, which turns out not to be easy given the profiles of the residents (Güzey, 2009; İslam, 2009; Uysal, 2012).

Roma families who went to Taşoluk had difficulties in paying the high rents and bills. The different socio-cultural profile of Taşoluk also put the displaced Sulukule residents in an even more difficult position. Those working in the entertainment industry in Sulukule lost their jobs due to transportation difficulties. Those who tend to engage in illegal activities due to unemployment and poverty made the other residents of Taşoluk uneasy. It was also a problem for the Roma, who lived in Sulukule their entire lives, to adapt to the lifestyle in mass houses. The lack of social solidarity in the neighbourhood and the obstacles to the continuation of ethno-cultural practices caused discontent among many Sulukule residents and they preferred to move to areas adjacent to Sulukule. Those who did not accept the impositions of the project and chose to stay in Sulukule had to struggle with the worsening living conditions. Along with the problems brought about by the demolitions, electricity and water cuts have become a threat to residents' health in the neighbourhood. Demolitions in Sulukule had been completed by September 2009 (Uysal, 2012).

The urban transformation process in Sulukule paved the way for the emergence of an unprecedented activist movement in the city. Urban activists organized and established a platform on the grounds that the transformation process in Sulukule was against the interests of the people of the region, especially the Roma population, and would impact the cultural diversity of the city (Sulukule Platform). With the efforts of this platform, the Sulukule Project has occupied an important place on the agenda of the city, the country, and even the world for more than three years. However, these efforts by the activists did not prevent the municipality from making agreements with the landlords and tenants in the area and demolishing almost all of the houses (İslam, 2009).

In Sulukule, despite serious objections, demolitions took place and the inhabitants were evacuated. The project, which was announced with the claim of improving the living conditions of the people living in the area, largely served to distribute them throughout the city. This result has been achieved despite the fact that the project is carried out only by public bodies without the involvement of the private sector and large public resources are used to mitigate the effects of the project (İslam, 2009). The urban poor who already faced challenging social and economic conditions were confronted with difficulties imposed by the diverse impacts of the urban transformation project. Many people living there left their neighbourhoods after selling their properties very cheaply as a result of the pressure and intimidation they faced. A large proportion of those transferred to TOKİ mass houses in Taşoluk have failed to make the necessary payments, leaving them in enormous financial debt to the state. On the other hand, those who cannot afford the housing provided by TOKİ are left to find solutions to their housing problem on their own in neighbourhoods where they have family, relatives, or other connections (Ünsal, 2013).

Conclusion

After the 2001 economic crisis, parallel with policies supporting the construction industry, urban transformation projects have accelerated and occupied an important place on the agenda of the government in Turkey. These projects also promoted by several pieces of legislation, served the appropriation of existing land for the use of high-income groups, giving rise to state-led gentrification.

When the legal regulations introduced in the 2000s regarding urban transformation are examined, it is revealed that the concerns about determining physical boundaries and building are at the forefront. No clue, foresight, or implementation system is included in any of these laws on issues such as how a certain settlement area will be transformed, how to prevent the inhabitants from being adversely affected by this transformation, or how to ensure that the economic and social order is not disturbed but is improved in the event of displacement of the inhabitants.

The Sulukule urban transformation project is one of the many cases resulting in state-led gentrification. These implementations bring with them negativities for the urban space and the population living in the transformed areas. Although relatively better urban spaces are created with the said transformations, no societal benefits are obtained. As a result of the exclusivity that has become a part of the urban transformation projects, low-income groups are forced to live at the periphery of the city after being displaced from many central neighbourhoods of the city, while high-income groups settle in these transformed areas and gain an advantageous position. While this situation causes significant segregation in the urban space, it also brings with it social polarization. Therefore, spatial and social inequalities in transforming areas emerge as an important problem area caused by this state-led gentrification.

References

Aalbers MB (2019) Introduction to the forum: From third to fifth-wave gentrification. *Tijdschrift voor Economische en Sociale Geografie* 110(1): 1–11.

Atkinson R and Bridge G (eds) (2005) *Gentrification in a Global Context: The New Urban Colonialism*. London: Routledge.

Crump J (2002) Deconcentration by demolition: Public housing, poverty, and urban policy. *Environment and Planning D: Society and Space* 20(5): 581–596.

Çavdar A and Tan P (ed) (2013) *İstanbul: Müstesna Şehrin İstisna Hali*. İstanbul: Sel Yayıncılık.

Ergün N (2006) Gentrification kuramlarının İstanbul'a uygulanabilirliği. In:Behar D and İslam T (eds) *İstanbul'da Soylulaştırma: Eski Kentin Yeni Sahipleri*. İstanbul: Bilgi Üniversitesi Yayınları,15–31.

Glass R (1964) Aspects of Change. In: Centre for Urban Studies (ed) *London Aspects of Change*. London: MacGibbon & Kee, xviii–xix.

Gordon R, Collins FL and Kearns R (2017) "It is the people that have made Glen Innes": State-led gentrification and the reconfiguration of urban life in Auckland. *International Journal of Urban and Regional Research* 41(5): 767–785.

Güzey Ö (2009) Sulukule'de kentsel dönüşüm devlet eliyle soylulaştırma. *Mimarlık*. 346: 73–80.

Hackworth J (2002) Postrecession gentrification in New York City. *Urban Affairs Review* 37(6): 815–843.

Hackworth J and Smith N (2001) The changing state of gentrification. *Tijdschrift voor Economische en Sociale Geografie* 92(4): 464–477.

İslam T (2009) Devlet eksenli soylulaşma ve yerel halk: Neslişah ve Hatice Sultan mahalleleri (Sulukule) örneği. Unpublished doctoral thesis. Yıldız Technical University.

Korkmaz C (2015) Evaluation of sustainability performance of urban regeneration projects: The case of the north entrance of Ankara urban regeneration project. Unpublished master's thesis. Middle East Technical University.

Lees L and Ley D (2008) Introduction to special issue on gentrification and public policy. *Urban Studies* 45(12): 2379–2384.

Lees L, Slater T and Wyly E (2008) *Gentrification*. New York and London: Routledge.

Özdemir D (ed) (2010) *Kentsel Dönüşümde Politika, Mevzuat, Uygulama: Avrupa Deneyimi, İstanbul Uygulamaları*. Ankara: Nobel Yayıncılık.

Slater T (2004) Municipally-managed gentrification in South Parkdale, Toronto. *The Canadian Geographer* 48(3): 303–325.

Slater T (2005) Gentrification in Canada's cities: From social mix to social "tectonics". In: R Atkinson and G Bridge (eds) *Gentrification in a Global Context: The New Urban Colonialism*. London: Routledge, 40–57.

Slater T (2006) The eviction of critical perspectives from gentrification research. *International Journal of Urban and Regional Research* 30(4): 737–757.

Slater T (2009) Missing Marcuse: On gentrification and displacement. *City* 13(2–3): 292–311.

Smith N (2002) New globalism, new urbanism: Gentrification as global urban strategy. *Antipode* 34: 427–450.

Sönmez F and Geniş Ş (2013) Türkiye soylulaştırma yazınının eleştirel bir değerlendirmesi. In: Tuna M (ed) *VII. Ulusal Sosyoloji Kongresi Bildiri Kitabı I*. -Muğla: Muğla Sıtkı Koçman Üniversitesi, 127–140.

Türkün A (ed) (2014) *Kentsel Ayrışmanın Son Aşaması Olarak Kentsel Dönüşüm. Mülk, Mahal, İnsan: İstanbul'da Kentsel Dönüşüm*. İstanbul: İstanbul Bilgi Üniversitesi Yayınları.

Türkün A and Sarıoğlu A (2014) Tarlabaşı: tarihi kent merkezinde yoksulluk ve dışlanan kesimler üzerinden yeni bir tarih yazılıyor. In: Türkün A (ed) *Mülk, Mahal, İnsan: İstanbul'da Kentsel Dönüşüm*. Istanbul: Istanbul Bilgi Üniversitesi Yayınları, 267–307.

Uysal ÜE (2012) Sulukule: Kentsel dönüşüme etno-kültürel bir direniş. *İdealkent* 3(7): 136–159.

Uzun CN (2001) *Gentrification in Istanbul: A Diagnostic Study*. Utrecht: KNAG.

Uzun N (2015) İstanbul'da şeçkinleştirme'nin üç aşaması: Cihangir, Galata ve Tarlabaşı üzerinden bir değerlendirme. In: Duman B and Coşkun İ (eds) *Neden, Nasıl ve Kim için Kentsel Dönüşüm*. İstanbul: Litera Yayıncılık, 431–451.

Uzun N (2020) Residential transformation leading to gentrification: Cases from Istanbul. In: Krase J and DeSena JN (eds) *Gentrification around the World, Volume I*. Cham: Palgrave Macmillan, 223–243.

Ünsal Ö (2013) *Innercity regeneration and the politics of resistance in Istanbul: A comparative analysis of Sulukule and Tarlabaşı*. Unpublished doctoral thesis. City University of London.

Van Weesep J (1994) Gentrification as a research frontier. *Progress in Human Geography* 18(1): 74–83.

Yüksel O (2007) Kentsel dönüşümün fiziksel ve sosyal mekâna etkisi: Kuzey Ankara girişi kentsel dönüşüm projesi. Unpublished master's thesis. Gazi University.

Index

Printed in the United States
by Baker & Taylor Publisher Services